Non-Recursive Behavioural Mo in Control Analy and Design

Non-Recursive Behavioural Models in Control Analysis and Design

By

Mihail Voicu and Octavian Pastravanu

Cambridge Scholars Publishing

Non-Recursive Behavioural Models in Control Analysis and Design

By Mihail Voicu and Octavian Pastravanu

Reviewers:
Viorel Barbu, full member of the Romanian Academy
Vlad Ionescu, corresponding member of the Romanian Academy

This book first published 2023

Cambridge Scholars Publishing

Lady Stephenson Library, Newcastle upon Tyne, NE6 2PA, UK

British Library Cataloguing in Publication Data
A catalogue record for this book is available from the British Library

ISBN (10): 1-5275-4296-3
ISBN (13): 978-1-5275-4296-9

CONTENTS

Part V. Experiments and Conclusions

PREFACE

This work develops a nonstandard approach to control systems analysis and design, to exploit the properties of a new type of behavioural model that has been called a *non-recursive* behavioural model in order to emphasize its differentiation from the *recursive* behavioural form of the classical state-space representations.

For a given real plant which exhibits a linear behaviour in the vicinity of any operating point, a non-recursive behavioural model is introduced as a coherent and consistent collection of discrete-time signals. These signals are acquired by sampling a certain number of appropriately selected input-state transfers of the plant, which, for a given time horizon, is actuated by piecewise constant input vectors. Thus a non-recursive behavioural model is defined as a pair of numerical matrices, built only on an experimental basis as follows: (i) a non-singular square matrix associated with the input evolution, each column of which is sequencing the piecewise constant input vectors that generate one of the selected input-state transfers; and (ii) a rectangular matrix associated with the forced response, whose columns are the final state vectors corresponding to the aforementioned time horizon and the selected input-state transfers.

If the input vectors are set to zero after this time horizon, further sampling of the selected input-state transfers, for a given number of periods, allows a set of rectangular matrices to be built associated with the free response. An extended non-recursive behavioural model is obtained by adding this set of matrices to the non-recursive behavioural model.

In terms of this construction, usual topics in linear control theory (such as controllability analysis, pole placement, structurally stable synthesis, linear quadratic control) can be addressed without any knowledge of the recursive behavioural model. Consequently, the need for a separate identification procedure is circumvented. Moreover, this approach has given deeper insight into the role played by the state-feedback mechanism, to make it possible to simplify (to some extent) the design algorithms. Results that have been derived within this novel framework demonstrate that non-recursive behavioural models contain the essential information required by system analysis and design. Besides the practical aspects that are directly related to self-learning control, the usage of non-recursive behavioural models has also materialized in substantial theoretical developments.

Initiated in the late-1970s by J. C. Willems (see Polderman and Willems (1998)), the *behavioural mathematical modelling* of dynamic systems was meant to solve inconsistencies in approaches based on state-space, transfer function, and convolution representation. A short introduction to the *behavioural mathematical modelling* of dynamic systems together with a clear distinction between non-recursive *behavioural* models and recursive *behavioural* models, on one side, and non-recursive models and recursive models, on the other (the last two introduced and used in social and human sciences), will be presented in Appendix A at the end of this book.

This book presents the contributions of the authors to create a unified scenario for control system analysis and design, exclusively based on *non-recursive behavioural models*. A large body of research (which has already been reported in several journal and conference papers and a PhD thesis authored by O. Pastravanu, and supervised by Mihail Voicu) has formed the background of this work which is divided into five parts and successively presents: some mathematical preliminaries and definitions of linear non-recursive behavioural models (Part I), techniques for state controllability analysis (Part II), procedures for feedback control design (Part III), optimal control design (Part IV) and some experimental results and conclusions (Part V).

A chapter-by-chapter description will certainly give more details about the addressed topics. Starting from a trajectory related geometric background of Ackermann's algebraic approaches to pole assignment based on linear recursive behavioural models (Chapter 1), non-recursive behavioural models are introduced in a natural manner (Chapter 2). Containing the same information as the recursive behavioural models, the non-recursive ones are used for the controllability analysis (Chapter 3) and for the evaluation of reachability sets for energy/amplitude constrained control (Chapter 4). Pertaining to the geometric background of pole assignment (leading to the definition of non-recursive behavioural models), these behavioural models are exploited to develop procedures for state feedback (Chapter 5) and structurally stable (Chapter 6) controller design. On the same basis, the linear quadratic problem can be solved in cases of open-loop control (Chapter 7) and closed-loop control (Chapter 8). The theoretical results are illustrated by simulation experiments (Chapter 9) and real-time experiments performed on a laboratory behavioural model (Chapter 10). In Chapter 11, conclusions are presented. Another two appendices, B and C, present some mathematical results used in Chapter 4. Finally, an extensive reference list is provided, with many quoted throughout the text, but no attempt has been made to reach any degree of completeness.

The book is equally useful to postgraduate students and research workers and practitioners interested in systems theory and its applications in various fields. Besides revealing a new and comprehensive interpretation of the basic principles that govern the behaviour of dynamical systems, the exploration of non-recursive behavioural models represents substantial potential for further research in automatic control, applied mathematics and related areas.

March 2023

The authors

ACKNOWLEDGEMENTS

Mihail Voicu is greatly indebted to the Commission of the European Union for the generous support he received as a visiting researcher at the Institute of Technology in Karlsruhe, when he prepared a preliminary version of this book. Special gratitude should be expressed to Prof. Dr.-Ing. Volker Krebs who, on accepting the research period, actually encouraged the authors' idea of writing the present book.

Mrs. S. L. Morecroft, from York (England), did a very professional language proofreading, and for her wonderful work the authors extend warm thanks to this very kind lady.

LIST OF FIGURES

NOTATIONS, SYMBOLS AND ACRONYMS

$\mathbf{N}$ – set of natural numbers

$\mathbf{N}_0 = \mathbf{N} \cup \{0\}$

$\mathbf{Q}$ – set of rational numbers

$\overline{1,n}$ – set of first n natural numbers

$\mathbf{Z}$ – set of integer numbers

$\mathbf{R}$ – set of real numbers

$\mathbf{R}_+$ – set of nonnegative real numbers

$\mathbf{C}$ – set of complex numbers

$\mathbf{R}^n$ – real vector space of order n

$\mathbf{R}^{m\times n}$ – set of $m \times n$ real matrices

$\mathrm{Span}_{\mathbf{R}}(x_1 \ldots x_k)$ – set of all linear combinations of the vectors x_i, $i = \overline{1,k}$, with coefficients in $\mathbf{R}$

$\mathrm{Pr}_p\, v$ – orthogonal projection of $v \in \mathbf{R}^n$ along $p \in \mathbf{R}^n$

$(\bullet)^\tau$ – transpose of $\bullet$ (vector or matrix)

$\|\bullet\|_l$ – l-norm of $\bullet$ (vector or matrix)

$\sigma(A)$ – spectrum of square matrix A

$\lambda_i(A)$ – i-th eigenvalue of square matrix A

$\mathrm{Tr}\,(A)$ – trace of square matrix A

$\det A$ – determinant of square matrix A

A^{-1} – inverse of square matrix A

A^+ – Moore–Penrose pseudoinverse of matrix A

$\mathrm{diag}\{\alpha_1 \ldots \alpha_n\}$ – diagonal matrix with entries $\alpha_i, i = \overline{1,n}$

$\mathrm{row}_i A$ – i-th row of matrix A

$\sigma(t)$ – Heaviside's (unit step) function

$[a_1, a_2, ..., a_n]$ – partitioning of a matrix into its columns $a_1, a_2, ..., a_n$

$\begin{bmatrix} b_1 \\ \vdots \\ b_n \end{bmatrix}$ – partitioning of a matrix into its rows $b_1, ..., b_n$

$[A_1, ..., A_n]$ – partitioning of a matrix into n column blocks

$\begin{bmatrix} B_1 \\ \vdots \\ B_n \end{bmatrix}$ – partitioning of a matrix into n row blocks

$[A]_{(i)}$ – i-th row block of matrix A

$\begin{bmatrix} A & B \\ C & D \end{bmatrix}$ – partitioning of a matrix into blocks A, B, C, D

I_n – unit matrix of order n

$\min_\alpha$ – minimum with respect to α

$\max_\alpha$ – maximum with respect to α

(X_j, U_j) – linear non-recursive behavioural model in j steps (see Definition 2.1)

$(X_{j+l}, ...X_j, U_j)$ – linear extended non-recursive behavioural control in $j+l$ steps (see Definition 2.3).

Acronyms

MEAC – measurement accuracy

COAC – computing accuracy

AC – global accuracy

$\mathrm{cod}\, S$ – controllability degree of system S

SVD – singular value decomposition

PART I.

INTRODUCTION TO LINEAR NON-RECURSIVE BEHAVIOURAL MODELS

CHAPTER 1

PRELIMINARIES BASED ON LINEAR RECURSIVE BEHAVIOURAL MODELS

1.1. Changing dynamics of control systems by pole placement

For a given real plant (Σ), assume that the basic operation principles exhibit the following three behavioural properties:

(i) The plant may be described by a state-space model and the state variables, denoted by $x \in \mathbf{R}^n$ are accessible to measurement

(ii) The plant is globally (asymptotically) stable

(iii) For small deviations from any operating point, the plant may be considered as linear, and there exists a minimal state-space realization expressed in terms of x:

$$\begin{aligned} \dot{x} &= A(\theta)x + B(\theta)u, \quad && t \in \mathbf{R}_+, x \in \mathbf{R}^n, u \in \mathbf{R}^m, \\ y &= C(\theta)x, && y \in \mathbf{R}^p. \end{aligned} \tag{1.1}$$

The numerical values of the entries of $A(\theta), B(\theta), C(\theta)$ are unknown and depend on the physical parameter vector θ of the current operating point.

Actually (1.1) represents a family of parameterized linear *recursive behavioural models*, denoted $S(\theta)$.

It is worth noticing that such situations are frequently encountered in fluid level, fluid pressure and temperature control. The behavioural model (1.1) is called *recursive* because it expresses the dynamical tendencies of the system evolution and not the system evolution itself. Comparatively, the solution of (1.1) for the initial condition $x(t_0)=x_0$, i.e.:

$$y(t)=C(\theta)e^{A(\theta)(t-t_0)}x_0+C(\theta)\int_{t_0}^{t}e^{A(\theta)(t-\tau)}B(\theta)u(\tau)d\tau,\quad t\geq t_0\geq 0,$$

expresses the actual evolution of the system. Notice that this solution represents a *non-recursive behavioural model*. As a matter of fact, the definition of dynamic systems (as a mathematical entity) was based from the very beginning by Kalman, see the book of Kalman et al. (1969) on *non-recursive* (as we say today) *behavioural* (a term introduced in the late-1970s by Willems, see the book of Polderman and Willems, (1998)) *models*, but updated to our approaches.

It should be noticed here that in the early-1980s, in the domains of social and human sciences (sociology, economy and psychology), the notions of recursive and non-recursive models were introduced as follows: (a) A recursive model is one where all causal effects are unidirectional and disturbances are uncorrelated; and (b) A non-recursive model contains one or more feedback loops or reciprocal causal effects and correlated disturbances. The reader can use Appendix A for a clear distinction between the *recursive* and *non-recursive behavioural models*, addressed in this book, and the *recursive models* and *non-recursive models*.

Consider the following two problems, which are now classical in control engineering:

(P) Design a state feedback controller to ensure that the closed-loop system performance is invariant with changes in the parameter vector θ.

(P ') When $p=m$, design a structurally stable discrete-time compensator for an exogenous step to ensure that the closed-loop system performance is invariant relative to changes in the parameter vector θ.

From the point of view of the digital controller design one has to take into account the fact that the parameter dependent matrices $A(\theta),B(\theta),C(\theta)$ in the linear recursive behavioural model (1.1), or $E(\theta),H(\theta)$ in its discrete-time pendant, with $T > 0$:

$$\begin{aligned} x(t+T) &= E(\theta)x(t) + H(\theta)u(t),\ t = kT,\ k \in \mathbf{N}, \\ y(t) &= C(\theta)x(t) \end{aligned} \tag{1.2}$$

are known with given accuracy, or they are partially or even completely unknown. In the last two cases the regular framework used for solving problems $(\mathbf{P})$ or $(\mathbf{P}')$ consists of two steps: a real-time identification of the discrete-time linear recursive behavioural model (1.2) corresponding to the actual operating point, followed by the synthesis of the controller relying on this recursive behavioural model.

For problem $(\mathbf{P})$ one considers the digital controller:

$$u(t) = F(\theta)x(kT),\ t \in [kT,(k+1)T],\ k \in \mathbf{N}, \tag{1.3}$$

and the gain matrix $F(\theta)$ is determined so the eigenvalues assigned to the closed-loop matrix $E(\theta) + H(\theta)F(\theta)$ will be preserved for any θ in the operating range.

For problem $(\mathbf{P}')$ one considers the augmented discrete-time linear behavioural model (S_θ') of the plant:

$$\begin{aligned} \underline{x}(t+T) &= \underline{E}(\theta)\underline{x}(t) + \underline{H}(\theta)u(t),\ t \in [kT,(k+1)T],\ k \in \mathbf{N}, \\ \underline{y}(t) &= \underline{C}(\theta)\underline{x}(t), \end{aligned} \tag{1.2'}$$

where $\underline{x}(t)$ includes the state variables $x_e \in \mathbf{R}^m$ of the internal recursive behavioural model associated with the exogenous variable according to Wonham (1979):

$$x_e(t+T) = x_e(t) - y(t) + r(t), \quad t = kT,\ k \in \mathbf{N},$$

and $r \in \mathbf{R}^m$ denotes the set point vector.

The gain matrix of the digital controller:

$$u(t) = \underline{F}(\theta)\underline{x}(kT), \quad t \in [kT, (k+1)T], \; k \in \mathbf{N}, \qquad (1.3')$$

should be determined so that the eigenvalues assigned to the closed-loop matrix $\underline{E}(\theta) + \underline{H}(\theta)\underline{F}(\theta)$ will be preserved for any θ in the operating range.

1.2. Implications of behavioural model accuracy

The quasi-totality of the approaches devoted to the problem $(\mathbf{P})$ or $(\mathbf{P'})$ offers purely algebraic solutions according to Kautsky (1985), Fletcher (1985) and Varga (1981), based on well-conditioned transforms applied to the matrices defining the linear recursive behavioural models (1.2) or (1.2'). Now it is important to notice that prescribing the closed-loop dynamics for the problem $(\mathbf{P})$ or $(\mathbf{P'})$ (e.g., by an optimality criterion) is theoretically equivalent to assigning the eigenvalues. Unfortunately, this equivalence represents a useful computational tool only when an accurate behavioural mathematical model is available. Measurement and computational errors result in significant differences between the spectrum actually assigned and the pre-specified one, obviously altering the real dynamics of the closed-loop system.

Taking into account the above-mentioned shortcomings of the usual framework for solving either $(\mathbf{P})$ or $(\mathbf{P'})$ and the fact that the main interest in synthesis is the desired closed-loop evolution, the algorithm used for the computation of the gain matrix in (1.3) (or (1.3')) should express a relation between the gain matrix and the state-space trajectory of (1.2) (or (1.2')), rather than a relation between the gain matrix and the matrices defining the linear recursive behavioural model (1.2) (or (1.2')). Such a direct approach is also suggested by Ackermann's multivariable assignment procedure of Ackermann (1977) (generally – when used only as an algebraic result and sometimes even contested – due to the necessity of inverting ill-conditioned matrices) which, unlike previously cited methods, is able to point out the role played by system controllability in achieving the desired closed-loop dynamics.

In this respect, a more profitable solution for **(P)** or **(P')** seems to be the usage of a well-organized and meaningful amount of input-state data (even when A, B or E, H are known), directly acquired from the functioning of the plant (Σ) and prepared as a non-recursive (and non-parametric) behavioural model which avoids the drawbacks of operating with the recursive behavioural model (1.3) (or (1.3')) and gives the possibility of reformulating the controller synthesis as a self-learning problem.

1.3. A trajectory-related geometric insight into multi-variable state feedback control

This section is devoted to the investigation of the geometric aspects of pole assignment to open the perspectives to a direct approach to controller design based on linear non-recursive behavioural models, which are rigorously defined in Chapter 2.

In order to simplify the mathematical notations, let us introduce the following linear constant recursive behavioural models of a real plant (Σ) corresponding to small deviations from a physical parameter vector θ *:

• the continuous-time linear recursive behavioural model:

$$\begin{aligned} &\dot{x} = Ax + Bu, \quad t \in \mathbf{R}_{+,} \;\; x \in \mathbf{R}^n, \;\; u \in \mathbf{R}^m, \\ &y = Cx, \qquad\qquad y \in \mathbf{R}^p, \end{aligned} \tag{1.4}$$

where $A = A(\theta\,*), B = B(\theta\,*), C = C(\theta\,*)$;

• the discrete-time linear recursive behavioural model:

$$\begin{aligned} x(t+T) &= Ex(t) + Hu(t), \;\; t = kT, \;\; k \in \mathbf{N}_0, \\ y(t) &= C\,x(t), \end{aligned} \tag{1.5}$$

where $T > 0$ and

$$E = E(\theta^*) = e^{A(\theta^*)T}, \quad H = H(\theta^*) = \int_0^T e^{A(\theta^*)t} dt B(\theta^*).$$

Furthermore, assume that the solution of problem $(\mathbf{P})$ for the recursive behavioural model (1.5) is given by the discrete-time linear state feedback controller:

$$u(t) \triangleq u_k = Fx(kT), \ t \in [kT, (k+1)T), \ k \in \mathbf{N}_0, \quad (1.6)$$

where $F = F(\theta^*)$.

The results will be expressed in terms of the constant recursive behavioural models (1.4), (1.5) and the constant controller (1.6) (corresponding to the operating point θ^*) and the principle of the method remains valid for any other operating point of a real plant (Σ).

The goal of this approach is to underscore the geometric background of pole assignment by presenting a trajectory related point of view issued as a necessity to complement the standard algebraic approach. This offers a "geometric-physical meaning" to the closed-loop dynamics achieved by a discrete-time linear state feedback controller. The results are presented in terms of the continuous-time linear recursive behavioural model (1.4) whose behaviour is to be modified by the discrete-time linear state feedback controller (1.6) in order to ensure that the closed-loop performance corresponds to the assigned spectrum:

$$\sigma(E + HF) = Z \triangleq \{z_i \in \mathbf{C}, \ i = \overline{1,n}; |z_i| < 1\}. \quad (1.7)$$

An arbitrary state trajectory $x(t)$ of the recursive behavioural model (1.4), corresponding to a non-zero initial state condition, can be studied by means of the decomposition:

$$x(t) = x_{fr}(t) + x_{fo}(t), \quad t \geq t_0 \geq 0, \quad (1.8)$$

where $x_{fr}(t)$, $x_{fo}(t)$ denote the free response and the forced response respectively, and t_0 is the initial instant. The relation between the two components (which from the physical point of view may be evidenced separately) represents the essence of the state feedback mechanism and will lead, in the next chapter, to a natural introduction of the linear non-recursive behavioural models.

When the control (1.6) is used in order to modify the dynamics of the recursive behavioural model (1.4), the corresponding system is actuated by a staircase signal and the algebraization induced by the recursive behavioural model (1.5) simplifies the mathematical description of the state trajectory of the behavioural model (1.4). In this respect the hyperplanes Π_i, $i=\overline{1,m}$, are introduced as follows:

$$\Pi_1 = \mathrm{Span}_{\mathbf{R}}(h_1, Eh_1, ..., E^{n_1-2}h_1, ..., h_m, Eh_m, ..., E^{n_m-1}h_m)$$

$$\cdots\cdots$$

$$\Pi_m = \mathrm{Span}_{\mathbf{R}}(h_1, Eh_1, ..., E^{n_1-1}h_1, ..., h_m, Eh_m, ..., E^{n_m-2}h_m),$$

where $h_i, i=\overline{1,m}$, denote the columns of H, and n_i, $i=\overline{1,m}$, denote the corresponding Kronecker invariants or structural indices of the discretized behavioural model (1.5). The geometric significance of Π_i, $i=\overline{1,m}$, concerning the recursive behavioural model (1.4) is given by the following result.

Proposition 1.1

The forced response of the recursive behavioural model (1.4) for any piecewise constant input with the sampling period $T>0$ is characterized by:

$$x_{fo}(t_0 + jT) \in \Pi_i, \; i=\overline{1,m}, \; j=\overline{0,n_i-1}, \; n \geq 2, \qquad (1.9)$$

for any initial instant $t_0 \geq 0$.

Proof: Since:

$$x_{fo}(t_0 + jT) = \sum_{k=1}^{j} E^{j-k} H u_{k-1}, \quad j = \overline{1, n_i - 1}, \qquad (1.10)$$

let us show that for any i, $i = \overline{1, m}$, the forced response $x_{fo}(t_0 + jT)$, $j = \overline{1, n_i - 1}$, can be expressed using the elements of the vector basis:

$$h_1, Eh_1, ..., E^{n_1 - 1} h_1, ..., h_m, Eh_m, ..., E^{n_m - 1} h_m,$$

with a vanishing component corresponding to the vector $E^{n_i - 1} h_i$.

The development of each term in (1.10) generally requires the basis vectors $E^k h_i, ..., E^k h_m, k = \overline{0, n_i - 2}$. On the other hand, the vector $E^{n_i - 1} h_i$ involves neither $E^k h_i$ with $k \leq n_i - 2$, nor $E^k h_l, l = \overline{1, m}, l \neq i$, $k = \overline{0, n_l - 2}$, for $n_l < n_i$. ■

Denote by p_i, $i = \overline{1, m}$, the unity vector of the orthogonal direction on Π_i, $i = \overline{1, m}$, and by $\mathrm{Pr}_{p_i} w$ the orthogonal projection of $w \in \mathbf{R}^n$ along p_i. Consider the vectors g_i, $i = \overline{1, m}$, satisfying:

$$g_i = \varepsilon_i p_i, \ \varepsilon_i \in \mathbf{R}, \ \text{ with } \ \varepsilon_i \, \mathrm{Pr}_{p_i} (E^{n_i - 1} h_i) = 1. \qquad (1.11)$$

The closed-loop system (1.4), (1.6) (i.e., described by (1.4), (1.6)) and its (free) response can be characterized by the following result.

Proposition 1.2

For any set of polynomials:

$$P_i(z) = \sum_{j=0}^{n_i} \alpha_{ij} z^j, \ i = \overline{1, m}, \ \text{ with } \ \alpha_{in_i} = 1. \qquad (1.12)$$

there exists $F \in \mathbf{R}^{m\times n}$ so that the closed-loop dynamics of the system (1.4), (1.6) satisfies the following equalities:

$$\sum_{j=0}^{n_i} \alpha_{ij} \mathrm{Pr}_{p_i} x(t_0 + jT) = 0, \; i = \overline{1,m}, \tag{1.13}$$

for any initial condition $x(t_0)$, $t_0 \geq 0$.

Proof: Consider the matrix of β -parameters of the pair (E,H)

$$M = \begin{bmatrix} 1 & \beta_{21} & \cdots & \beta_{i1} & \cdots & \beta_{m1} \\ 0 & 1 & & \beta_{i2} & \cdots & \beta_{m2} \\ & 0 & & \vdots & & \\ \vdots & & & \beta_{ii-1} & & \vdots \\ & & & 1 & & \\ \vdots & & & 0 & & \vdots \\ & & & & & \beta_{mm-1} \\ 0 & & \cdots & & 0 & 1 \end{bmatrix}. \tag{1.14}$$

Since M is non-singular, F in state feedback control (1.6) may be uniquely written as:

$$F = MK, \quad K \in \mathbf{R}^{mxn}.$$

According to (Popov, 1972), matrix K exists so that the closed-loop discrete-time linear recursive behavioural model (1.5), (1.6) has the α-parameters defined by the matrix:

$$L_c = \left[\begin{array}{cccc|cccc|c|cccc} \alpha_{10} & \alpha_{11} & \cdots & \alpha_{1n_1-1} & 0 & 0 & \cdots & 0 & & 0 & 0 & \cdots & 0 \\ 0 & 0 & \cdots & 0 & \alpha_{20} & \alpha_{21} & \cdots & \alpha_{2n_2-1} & \cdots & 0 & 0 & \cdots & 0 \\ \vdots & \vdots & & \vdots & \vdots & \vdots & & \vdots & & \vdots & \vdots & & \vdots \\ 0 & 0 & \cdots & 0 & 0 & 0 & \cdots & 0 & & \alpha_{m0} & \alpha_{m1} & \cdots & \alpha_{mn_m-1} \end{array}\right]. \tag{1.16}$$

Using the non-singular linear transform defined by Luenberger (1967):

$$Q = \begin{bmatrix} Q_1 \\ Q_2 \\ \vdots \\ Q_m \end{bmatrix} \text{ with } Q_i = \begin{bmatrix} g_i^{\tau} \\ g_i^{\tau} E \\ \vdots \\ g_i^{\tau} E^{n_i - 1} \end{bmatrix}, \tag{1.17}$$

where τ denotes the transposition, the closed-loop matrix of the discrete-time system, according to Brunovski (1970), is:

$$\tilde{E}_c = Q(E - HMK)Q^{-1} = \begin{bmatrix} \tilde{E}_{c_{11}} & \cdots & 0 & \cdots & 0 \\ \vdots & & \vdots & & \vdots \\ 0 & \cdots & \tilde{E}_{c_{ii}} & \cdots & 0 \\ \vdots & & \vdots & & \vdots \\ 0 & \cdots & 0 & \cdots & \tilde{E}_{c_{mm}} \end{bmatrix}, \tag{1.18}$$

where

$$\tilde{E}_{c_{ii}} = \begin{bmatrix} 0 & 1 & 0 & \cdots & 0 \\ 0 & 0 & 1 & \cdots & 0 \\ \vdots & \vdots & \vdots & & \vdots \\ 0 & 0 & 0 & \cdots & 1 \\ -\alpha_{i0} & -\alpha_{i1} & -\alpha_{i2} & \cdots & -\alpha_{in_i-1} \end{bmatrix} \in \mathbf{R}^{n_i \times n_i}. \tag{1.19}$$

Noticing the block diagonal form of $\tilde{E}_c$ in (1.18), it results:

$$\sum\nolimits_{j=0}^{n_i} \alpha_{ij} \tilde{E}_c^j = \operatorname{diag}\left\{ \sum\nolimits_{j=0}^{n_i} \alpha_{ij} \tilde{E}_{c_{11}}^j, \ldots, \sum\nolimits_{j=0}^{n_i} \alpha_{ij} \tilde{E}_{c_{mm}}^j \right\},$$

and taking into account the companion form of $\tilde{E}_{c_{ii}}$ in (1.19), one can write:

$$\sum\nolimits_{j=0}^{n_i} \alpha_{ij} \tilde{E}_c^j = \operatorname{diag}\left\{ \sum\nolimits_{j=0}^{n_i} \alpha_{ij} \tilde{E}_{c_{11}}^j, \ldots, \underset{i}{0}, \ldots, \sum\nolimits_{j=0}^{n_i} \alpha_{ij} \tilde{E}_{c_{mm}}^j \right\}. \tag{1.21}$$

On the other hand:

$$\sum_{j=0}^{n_i} \alpha_{ij} \Pr_{p_i} x(t_0+jT) = p_i^\tau \left[\sum_{j=0}^{n_i} \alpha_{ij} (E-HMK)^j x(t_0) \right] =$$
$$= p_i^\tau Q^{-1} \sum_{j=0}^{n_i} \alpha_{ij} \tilde{E}_c^j Q x(t_0), \qquad (1.22)$$

with

$$p_i^\tau Q^{-1} = [\underbrace{0\ldots0}_{n_1}\ldots\underbrace{0\ldots0}_{n_i-1}\underbrace{10\ldots0}_{n_i}\ldots\underbrace{0\ldots0}_{n_m}], \quad i=\overline{1,m}.$$

Therefore, using (1.22), it is commonplace to see that (1.13) is fulfilled for $F = MK$. ■

Consider now the polynomial $P(z)$ whose zeros are the elements of Z in (1.7) which can be factorized as follows:

$$P(z) = \prod_{i=1}^{m} P_i(z) \qquad (1.23)$$

with $P_i(z)$, $i = \overline{1,m}$, defined by (1.12).

Proposition 1.3

For any $P(z)$ associated with the set Z defined (1.7) and factorized according to (1.23), there exists $F \in \mathbf{R}^{m\times n}$ which ensures, for the closed-loop behavioural model (1.4), (1.6), the pole placement into set (1.7), satisfying:

$$\Pr_{p_i} x_{fo}(t_0 + n_i T) = -\sum_{j=0}^{n_i} \alpha_{ij} \Pr_{p_i} x_{fr}(t_0 + jT),\ i = \overline{1,m}, \ (1.24)$$

for any initial condition $x(t_0)$, $t_0 \geq 0$.

Proof: Equalities (1.13) can be rewritten in the following manner:

$$\sum_{j=0}^{n_i} \alpha_{ij} \Pr_{p_i} x_{fo}(t_0 + jT) = -\sum_{j=0}^{n_i} \alpha_{ij} \Pr_{p_i} x_{fr}(t_0 + jT),\ i = \overline{1,m}. \qquad (1.25)$$

From Proposition 1.1, it follows that for all i's for which the structural indices n_i are satisfying $n_i \geq 2$, one may write:

$$\sum_{j=0}^{n_i-1} \alpha_{ij} \Pr_{p_i} x_{fo}(t_0 + jT) = 0.$$

In other words, for each $i = \overline{1,m}$, the equalities:

$$\sum_{j=0}^{n_i} \alpha_{ij} \Pr_{p_i} x_{fo}(t_0 + jT) = \Pr_{p_i} x_{fo}(t_0 + n_i T),\ i = \overline{1,m},$$

show that (1.25) is equivalent to (1.24) and therefore the existence of $F = MK$ is proved. ■

Remark 1.1

The relations between the free response and the forced response of (1.4) expressed by (1.24) show that the directions of p_i, $i = \overline{1,m}$, can be considered as geometric invariants of the system (1.4) actuated by a piecewise constant control with the sampling period $T > 0$. Obviously, these directions depend on the value of T. ■

The right-hand sides of equalities (1.24) are completely defined, on the one hand, by the free response of the system (1.4), and on the other, by the prescribed closed-loop dynamics for the system (1.4), (1.6). The only elements that take into account the influence of the gain matrix F on the closed-loop dynamics are $\Pr_{p_i} x_{fo}(t_0 + n_i T)$, $i = \overline{1,m}$. Further analysis of (1.24) will render evident the following result.

Proposition 1.4

For any polynomial $P(z)$ associated with Z given by (1.7) and factorized according to (1.23), there exists $F \in \mathbf{R}^{m \times n}$ which ensures, for the closed-loop system (1.4), (1.6), the pole placement into (1.7), satisfying:

$$\text{row}_i(M^{-1}F)x(t_0) = -\varepsilon_i \sum_{j=0}^{n_i} \alpha_{ij} \,\text{Pr}_{p_i} x_{fr}(t_0+jT), \; i=\overline{1,m}, \quad (1.26)$$

for any initial condition $x(t_0)$, $t_0 \geq 0$. By $\text{row}_i(M^{-1}F)$ one denotes the i -th row of matrix $(M^{-1}F)$ and $\varepsilon_i \in \mathbf{R}$ is defined according to (1.11).

Proof: First let us prove that the equalities:

$$\text{Pr}_{p_i}(E^{n_i-1}\tilde{h}_l) = 0, \; l=\overline{1,m}, \; l \neq i, \quad (1.27a)$$

hold for any i, $i=\overline{1,m}$, where $\tilde{h}_l$ denotes the l -th column of the matrix HM.

The Kronecker invariants of the pair (E, HM) being the same as those of the pair, according to Brunovski (1970), the vector set:

$$\tilde{h}_1, \ldots, E^{n_1-1}\tilde{h}_1, \ldots, \tilde{h}_m, \ldots, E^{n_m-1}\tilde{h}_m \quad (1.27b)$$

represents a vector basis in $\mathbf{R}^n$. Let n_l be the Kronecker invariants or structural indices corresponding to $\tilde{h}_l$.

In the case $n_l < n_i$ $(n_i \geq 2)$, from the equality:

$$E^{n_i-1}\tilde{h}_l = E^{n_i-1-n_l}E^{n_l}\tilde{h}_l,$$

and by using the possibility of expressing $E^{n_l}\tilde{h}_l$ only in terms of $E^j h_k$, $k=\overline{1,m}$, $j=\overline{0,n_l-1}$, (as the β -parameters of the pair (E, HM) are zero (Ackermann, 1985)), $E^{n_i-1}\tilde{h}_l$ can be expressed only in terms of the vectors $E^j\tilde{h}_k$, $k=\overline{1,m}$, $j=\overline{0,n_i-2}$. In other words, $E^{n_i}\tilde{h}_l$ can be expressed using only the vectors $E^j h_k$, $k=\overline{1,m}$, $j=\overline{0,n_i-2}$, which proves (1.27a) for the above considered case.

For $n_l \geq n_i$, when $l < i$, $E^{n_l-1}\tilde{h}_l$ can be expressed using only the vectors $E^j\tilde{h}_k, k=\overline{1,l}, j=\overline{0,n_i-1}$, which proves (1.27a) in this case.

For $n_l \geq n_i$, when $l \geq i$, one has $\beta_{li} = 0$, i.e., $\tilde{h}_l$ does not depend on h_i and the vector $E^{n_l-1}\tilde{h}_l$ which belongs to the basis (1.27b) has no component along the vector $E^{n_i-1}h_i$. This completes the proof of (1.27).

Now since

$$\mathrm{Pr}_{p_i} x_{fo}(t_0 + n_i T) = \mathrm{Pr}_{p_i}(E^{n_i-1}HMKx(t_0)),\ i = \overline{1,m},$$

taking into account (1.27a), one can write:

$$\mathrm{Pr}_{p_i}[E^{n_i-1}HMK\,x(t_0)] = \mathrm{Pr}_{p_i}(E^{n_i-1}\tilde{h}_i)(\mathrm{row}_i K)x(t_0).$$

As

$$\tilde{h}_i = h_i + \sum_{k=1}^{i-1} \beta_{ik} h_k,$$

it follows that:

$$\mathrm{Pr}_{p_i}(E^{n_i-1}\tilde{h}_i) = \mathrm{Pr}_{p_i}(E^{n_i-1}h_i).$$

As a matter of fact, the proof is complete if one takes into consideration the role of matrix K introduced by (1.15). ■

Thus Proposition 1.4 opens the possibility of expressing the state feedback mechanism in terms of the open-loop dynamics of the system (1.4) in a very transparent manner.

Theorem 1.1

Consider $P(z)$ associated with Z, defined by (1.7) and factorized by

(1.23). Then the state feedback (1.6), which ensures the pole placement into set (1.7) for the closed-loop system (1.4), (1.6), is given by:

$$F = MK \tag{1.28}$$

where the matrix

$$K^{\tau} = [k_1, \ldots, k_m] \in \mathbf{R}^{n \times m} \tag{1.29}$$

is built with the unique solutions $k_i \in \mathbf{R}^n$ of the following m algebraic systems (with the same coefficients in the left-hand side):

$$[x^q(t_0^q)]^{\tau} k_i = -\varepsilon_i \sum\nolimits_{j=0}^{n_i} \alpha_{ij} \mathrm{Pr}_{p_i} x_{fr}^q (t_0^q + jT), \; q = \overline{1,n}, \; i = \overline{1,m}. \tag{1.30}$$

The initial conditions considered $x^q(t_0^q)$, $t_0^q \geq 0$, $q = \overline{1,n}$, are linearly independent, and $\varepsilon_i \in \mathbf{R}$, $i = \overline{1,m}$, are defined by (1.11).

Proof: For any i, $i = \overline{1,m}$, the statement (1.30) is a consequence of (1.25) written for n linearly independent initial conditions $x^q(t_0^q), t_0^q \geq 0$, $q = \overline{1,n}$. If one adds to (1.30), the supplementary equations of type (1.25), one obtains an overdetermined algebraic system with rank n, which satisfies the Kronecker–Capelli theorem. ■

If

$$n_{\delta} = \max_{i=\overline{1,m}} n_i$$

denotes the controllability index of the discrete-time system (1.5), then the controller (1.6) can be determined such that the minimal polynomial of the closed-loop system (1.4), (1.6) has the degree n_{δ}.

Corollary 1.1

A state feedback controller (1.6) ensuring the minimal polynomial $P_{\min}(z)$, (degree $P_{\min}(z)=n_\delta$) for the closed-loop system (1.4), (1.6), is given by (1.28)–(1.30) with $P_{\min}(z)$ factorized according to (1.23) and $P_\delta(z)=$ $=P_{\min}(z)$, $P_i(z)\,|\,P_\delta(z)$ for $i=\overline{1,m}$.

Proof: Using the Luenberger canonical form and the same notations and techniques as in the proof of Proposition 1.2, it is easy to show that:

$$\sum_{j=0}^{n_\delta}\alpha_{\delta j}\tilde{E}_c^j=0.$$

This means that in terms of detailed loop matrices, one can write:

$$\sum_{j=0}^{n_\delta}\alpha_{\delta j}(E-HMK)^j=0_{n\times n}. \quad ■$$

Remark 1.2

Theorem 1.1 can be derived directly as an algebraic consequence of Theorem 4 in Ackermann (1977) or of Theorem 3 in Ackermann (1984), but the development presented in this section has outlined the geometric implications of changing the dynamics of (1.4) by using a state feedback controller (1.6), responsible for the closed-loop eigenvalue location.

Taking into account the initial conditions:

$$x^q(t_0^q)=e_q=[0...0\underset{q}{1}0...0]^\tau,\ q=\overline{1,n},$$

in (1.30) one can note that the scalar value of the gain factor $K(i,q)$, $i=\overline{1,m}$, depends both on the evolution of the open-loop free response along p_i (i.e., $\sum_{j=0}^{n_i}\alpha_{ij}\mathrm{Pr}_{p_i}x_{fr}^q(t_0^q+jT)$, when the system (1.4) is initialized in a unit norm state and on the evolution of the open-loop forced response along p_i i.e., $\varepsilon_i=(\mathrm{Pr}_{p_i}(E^{n_i-1}h_i))^{-1}$ when the controller input has a unit norm. ■

When the free evolution is comparable with the forced one, the value of the gain factor k_i is of the unity order; otherwise, for a free evolution dominance (in comparison with the forced evolution) the value of the gain factor k_i, increases. At the same time, it is worth stressing the influence of the β -parameters expressed by M upon the gain matrix F. Even though these remarks concerning a qualitative interpretation of controllability in the context of eigenvalue assignment are not presented, as discussed by Ackermann (1977), the vectors g_i $i = \overline{1,m}$, used by the cited paper have been called "controllability vectors", which has already suggested the necessity of a thorough investigation in the geometry of the state-space trajectories.

Remark 1.3

Theorem 1.1 represents a natural generalization, according to Pastravanu (1992) and Pastravanu and Voicu (1992b), of a similar result derived for the pole assignment in a single-input case given by Pastravanu and Voicu (1992a). ■

This enriched understanding of the geometric background allows the modification of the closed-loop behaviour by means of the state feedback controls intimately related to the open-loop dynamics decomposed according to (1.8) and hints that the recursive behavioural mathematical model could be avoided in pole placement. Hence it could be replaced by a meaningful and well-organized collection of data directly acquired by measurement. In the following chapter these sets of experimentally acquired data will be organized as chronologically indexed sequences called non-recursive behavioural models and will represent the prerequisites for the design of a state feedback controller according to problem **(P)** or **(P')** stated in §1.1.

CHAPTER 2

LINEAR NON-RECURSIVE BEHAVIOURAL MODELS

The evolution of the recursive behavioural model (1.4) has already been expressed by (1.8), where x_{fo} denotes the forced response (under the action of u, with a zero initial condition) and x_{fr} denotes the free response (representing the system dynamics with a non-zero initial condition). These two components, which from the physical point of view may be evidenced separately, represent the starting point for introducing the discrete-time linear non-recursive behavioural model with the sampling and holding period $T > 0$, as a collection of data that completely describes the forced dynamics of a given system, and the free one as well, in a consistent and coherent manner. The usefulness of this collection of data will be proved step by step by replacing the recursive behavioural mathematical model in some important problems of system analysis and synthesis.

2.1. Non-recursive behavioural models of forced response

Using in (1.4), piecewise constant inputs:

$$u(t) = u_k,\ t \in [(k-1)T, kT),\ k \in \mathbf{N}, \qquad (2.1)$$

where

$$u_k = [u_{1k} \ldots u_{mk}]^{\tau}$$

for a zero initial condition, the response of the system (1.4) for $t = jT$ has the following form:

$$x_j \triangleq x(jT) = \sum_{k=1}^{j} E^{j-k} H u_{k-1},\ x_0 = 0,\ j = 1,2,..., \quad (2.2)$$

For the sake of simplicity let us express the sequence $u_0, u_1, ..., u_{j-1}$ by:

$$\overline{u}_j \triangleq [u_{j-1}^{\tau}, u_{j-2}^{\tau}, ..., u_1^{\tau}, u_0^{\tau}]^{\tau} \in \mathbf{R}^{mj}, \quad (2.3)$$

and the sequence $H, EH, ..., E^{j-1}H$ by:

$$R_j \triangleq [H, EH, ..., E^{j-1}H] \in \mathbf{R}^{n\times mj}, \quad (2.4)$$

which represents the controllability matrix in j steps associated with system (1.5).

From (2.2)–(2.4), it follows that the forced response of system (1.4) for $t = jT$ may be rewritten as follows:

$$x_j = R_j \overline{u}_j,\ x_0 = 0,\ j = 1,2,... . \quad (2.5)$$

When applying to the real system (Σ) a set of mj control sequences of length j (relation (2.3)), denoted by $\overline{u}_j^i,\ i = \overline{1, mj}$, the set of state vectors $x_j^i,\ i = \overline{1, mj}$, can be written as:

$$X_j = R_j U_j, \quad (2.6)$$

where

$$X_j \triangleq [x_j^1, ..., x_j^{mj}] \in \mathbf{R}^{n\times mj}, \quad (2.7)$$

$$U_j \triangleq [\overline{u}_j^1, ..., \overline{u}_j^{mj}] \in \mathbf{R}^{mj\times mj}. \quad (2.8)$$

Definition 2.1

The pair of matrices (X_j, U_j) defined by (2.7), (2.8), with U_j, being non-singular, represents a discrete-time non-recursive linear behavioural mathematical model in j steps of a plant (Σ), or simply a linear non-recursive behavioural model. ■

Remark 2.1

An infinite number of linear non-recursive behavioural models, satisfying Definition 2.1 are available, for each fixed j, and, their equivalence is ensured by the unicity of R_j in (2.6). ■

Remark 2.2

A discrete-time linear non-recursive behavioural model for any real plant (Σ) represents, in fact, a consistent and coherent collection of data (denoted by (X_n, U_n) in Definition 2.1) directly acquired by measurements during the functioning of the plant (Σ) without any computational processing of data as it requires the identification of the linear recursive behavioural models (1.4) and (1.5). ■

2.2. Orthogonal test signals

Definition 2.2

The pair (X_j, U_j) in Definition 2.1 is said to be "a linear non-recursive behavioural model with orthogonal test signals" if matrix U_j has the form:

$$U_j = \alpha\Theta, \tag{2.9}$$

where $\alpha > 0$ and $\Theta \in \mathbf{R}^{mj \times mj}$ is an orthogonal matrix. ■

Constant α plays the part of a global amplitude, which is adjustable in order to benefit from the best accuracy of the measurements according to Remark 2.2 formulated by Voicu and Pastravanu (1987).

For the class of signals (2.9), the numerical problem of determining $\text{rank}X_j$ remains invariant with respect to the particular set of controls employed.

In the special case of the 2^k-th order systems, $k \in \mathbf{N}$, one can use the first 2^k Haar or Walsh functions according to Ahmed and Rao (1975). The latter option has the advantage of being in the easy to synthesize "bang-bang" form.

In the general case, the application of Householder factorization according to Golub and Van Loan (1983) to a square matrix with a uniform random number on the interval $(-1, 1)$ represents a source of obtaining orthogonal matrices.

2.3. Extended non-recursive behavioural models

For the investigation of the free response of a real plant (Σ), one considers for $t \in [jT,(j+l)T)$, $l \in \mathbf{N}$, zero test signals, i.e., for each test of index i, $i = \overline{1, mj}$, one adds:

$$u_{j+k-1} = [0,...,0]^\tau, \; 1 \le k \le l. \tag{2.10}$$

In this way, the linear non-recursive behavioural model will also incorporate information about the free dynamics of a real system (Σ) by means of the matrices:

$$X_{j+k} \triangleq [x^1_{j+k},...,x^{mj}_{j+k}] \in R^{n \times mj}, \; 1 \le k \le l. \tag{2.11}$$

For these matrices one can write the equations:

$$X_{j+k} = EX_{j+k-1}, \; 1 \le k \le l. \tag{2.12}$$

Definition 2.3

The set of $l+2$ matrices $(X_{j+l},...,X_j,U_j)$ defined by (2.7), (2.8) and

(2.11), with U_j being non-singular, represents an extended discrete-time non-recursive linear behavioural mathe-matical model in $j+l$ steps of a plant (Σ) or simply an extended linear non-recursive behavioural model. ■

Remark 2.3

The extended linear non-recursive behavioural model offers separate information about the forced and free responses, respectively. ■

Remark 2.4

The experimental determination of an extended linear non-recursive behavioural model needs test signals U_j which must ensure a satisfactory accuracy in acquiring the elements of matrices X_{j+k}, $k=\overline{0,l}$. ■

2.4. A non-recursive behavioural model derived from a step response

The following equivalence established between the extended non-recursive behavioural model and the step response of a plant (Σ) will especially justify the naturalness and usefulness of the extended linear non-recursive behavioural model.

Step signals of amplitude $\alpha>0$:

$$u_i(t)=\alpha\sigma_i(t),\ i=\overline{1,m},$$

with

$$\sigma_i(t)=[0...0\ \underset{i}{1}\ 0...0]^{\tau},\ t\in\mathbf{R}_+,\ i=\overline{1,m},$$

are successively applied to the input of a plant (Σ). Under the zero initial state condition, for each $u_i(t)$, $i=\overline{1,m}$, one determines, for a sampling period $T>0$, the first $j+l$ values of the step response denoted by:

$$x_{\sigma_i}(T), x_{\sigma_i}(2T), ..., x_{\sigma_i}((l+j)T),\ i=\overline{1,m}. \qquad (2.13)$$

One constructs, now, the matrix:

$$D=[d_1, d_2, ..., d_{j+l}] \in \mathbf{R}^{n\times m(j+l)} \qquad (2.14)$$

where the blocks d_r, $q=\overline{1,j+l}$, have the form:

$$d_1=[x_{\sigma_1}(T), ..., x_{\sigma_m}(T)] \in \mathbf{R}^{n\times m} \qquad (2.15)$$

and

$$d_q=[x_{\sigma_1}(qT)-x_{\sigma_1}((q-1)T), ..., x_{\sigma_m}(qT)-x_{\sigma_m}((q-1)T) \in \mathbf{R}^{n\times m}. \qquad (2.16)$$

One extracts from D the matrices:

$$D_k=[d_{k+1}, ..., d_{k+j}] \in \mathbf{R}^{n\times jm}, k=\overline{0,l}. \qquad (2.17)$$

Theorem 2.1

Matrices X_{j+k}, $k=\overline{0,l}$, of the extended linear non-recursive behavioural model in $j+k$ steps, according to Definition 2.3, where $U_j=\alpha I_{mj}$ (I_{mj} denoting the mj order unit matrix), $\alpha>0$, satisfy the equalities:

$$X_{j+k}=D_k,\ k=\overline{0,l}. \qquad (2.18)$$

Proof: With $U_j=\alpha I_{mj}$ in (2.6) and according to (2.4) it follows that:

$$X_{j+k}=\alpha E^k R_j,\ k=\overline{0,l}. \qquad (2.19)$$

On the other hand, denoting by h_i the i-th column of H, one can write:

$$x_{\sigma_i}(qT) = \alpha \sum_{p=1}^{r} E^{p-1} h_i, \; q = \overline{1, j+l}, \; i = \overline{1, m}, \quad (2.20)$$

which means that:

$$d_q = \alpha E^{r-1} H, \; q = \overline{1, j+l},$$

and

$$D_k = [\alpha E^k H, \alpha E^{k+1} H, ..., \alpha E^{k+j-1} H] = \alpha E^k R_j, \; k = \overline{0, l}, \quad (2.21)$$

respectively.

Now (2.18) results from (2.19) and (2.21). ■

Remark 2.5

As a matter of fact, Theorem 2.1 proves that a suitably chosen sequence of samples of the step response contains consistent and coherent information concerning both the free and forced behavioural response of a plant (Σ) according to its extended linear non-recursive behavioural model (Definition 2.3). ■

Remark 2.6

The simplicity of the tests based on step responses offers the advantage of experiment repeatability, to allow an operation with average values obtained after several experiments, which reduces the unavoidable noisy effects accompanying the measurements. ■

Remark 2.7

The step inputs $u_i(t)$, $i = \overline{1, m}$, may produce significant variations of the state variables, to ensure a satisfactory accuracy of measurements, under the condition that an adequate sampling and holding period $T > 0$ has been chosen. It must be noted here that short and long sampling periods can lead to less accurate entries in the matrices D_k and consequently to numerical drawbacks. ■

Remark 2.8

An extended linear non-recursive behavioural model may be obtained by a self-learning experiment devoted to the real plant (Σ). The usage of this experimental knowledge in solving problems **(P)** and **(P')** (see 1.1) circumvents the need for a linear recursive behavioural model as required by the usual methods. ■

2.5. A non-recursive behavioural model as a mathematical entity

In previous sections of this chapter, (extended) non-recursive behavioural models have been introduced starting from the dynamics of a system presumed to have a linear behaviour; moreover, it has been shown that for a discrete-time system (1.5) one can build an infinity of (extended) non-recursive behavioural models. In this context, the following two questions arise naturally:

- Given $n, m \in \mathbf{N}$, and two arbitrary matrices $X_j \in \mathbf{R}^{n \times mj}$, $U_j \in \mathbf{R}^{mj \times mj}$, with U_j being non-singular, what are the conditions ensuring that (X_j, U_j) is a non-recursive behavioural model, in accordance with Definition 2.1
- Given $n, m \in \mathbf{N}$, and an arbitrary set of $l+2$ matrices $X_{j+l} \in \mathbf{R}^{n \times mj}, \ldots, X_j \in \mathbf{R}^{n \times mj}, U_j \in \mathbf{R}^{mj \times mj}$, with U_j being non-singular, what are the conditions ensuring that $(X_{j+l}, \ldots, X_j, U_j)$ is an extended non-recursive behavioural model, in accordance with Definition 2.3?

The answers will be formulated by two theorems which characterize the non-recursive and extended non-recursive behavioural models as mathematical entities. Using the same notations as above, since U_j is non-singular, there exists a unique matrix:

$$\Gamma_j = X_j U_j^{-1},\ \Gamma_j \in \mathbf{R}^{n\times mj}. \qquad (2.22)$$

where partition Γ_j is as follows:

$$\Gamma_j = [\gamma_1,...,\gamma_j],\ \gamma_i \in \mathbf{R}^{n\times m},\ i=\overline{1,j}.$$

Theorem 2.2

The pair of matrices (X_j, U_j) is a non-recursive behavioural model in j steps if and only if:

$$\operatorname{rank}[\gamma_1,...,\gamma_{j-1}] = \operatorname{rank}\begin{bmatrix}\gamma_1,...,\gamma_{j-1}\\ \gamma_2,...,\gamma_j\end{bmatrix}. \qquad (2.24)$$

Proof: According to (2.6), the pair (X_j, U_j) is a non-recursive behavioural model *if* there exist two matrices $E\in \mathbf{R}^{n\times n}$, $H\in \mathbf{R}^{n\times m}$ such that:

$$\gamma_i = E^{i-1}H,\ i=\overline{1,j}. \qquad (2.25)$$

This condition is equivalent to

$$\begin{aligned}&\gamma_1 = H,\\ &\gamma_{i+1} = E\gamma_i,\ i=\overline{1,j-1},\end{aligned} \qquad (2.26)$$

which can be regarded as a linear algebraic system having solutions *if* (2.24) is fulfilled. ■

Theorem 2.3

The set of matrices $(X_{j+l},...,X_j,U_j)$ is an extended non-recursive behavioural model in $j+l$ steps if and only if:

$$\operatorname{rank}[\gamma_1,\ldots,\gamma_{j-1},X_j\ldots,X_{j+l-1}]=\operatorname{rank}\begin{bmatrix}\gamma_1,\ldots,\gamma_{j-1},X_j\ldots,X_{j+l-1}\\ \gamma_2,\ldots,\gamma_j,X_{j+1}\ldots,X_{j+l}\end{bmatrix} \quad (2.27)$$

Proof: According to (2.6) and (2.12), the set of matrices $(X_{j+l}, \ldots, X_j, U_j)$ is an extended non-recursive behavioural model *if* there exist two matrices $E \in \mathbf{R}^{n\times n}$, $H \in \mathbf{R}^{n\times m}$ such that:

$$\begin{aligned}&\gamma_i = E^{i-1}H,\ i=\overline{1,j},\\ &X_{j+k} = EX_{j+k-1}\ k=\overline{1,l}.\end{aligned} \quad (2.28)$$

This condition is equivalent to

$$\begin{aligned}&\gamma_1 = H,\\ &\gamma_{i+1} = E\gamma_i,\ i=\overline{1,j-1},\\ &X_{j+k} = EX_{j+k-1},\ k=\overline{1,l},\end{aligned} \quad (2.29)$$

which can be regarded as a linear algebraic system having solutions *if* (2.27) is fulfilled. ■

Remark 2.9

As seen from the proofs, these characterization theorems are exploiting the link between the non-recursive and the recursive behavioural models. Therefore, (2.24) can also be seen as a necessary and sufficient condition for the existence of at least one recursive behavioural model (1.5), such that (X_j, U_j) represents a non-recursive behavioural model built according to Definition 2.1. Similarly, (2.27) can also be seen as a necessary and sufficient condition for the existence of at least one recursive behavioural model (1.5), such that $(X_{j+l},\ldots,X_j,U_j)$ represents an extended non-recursive behavioural model built according to Definition 2.3. Note that the conditions which ensure the unicity will be separately explored in the next chapter. ■

Remark 2.10

Given $m, n \in \mathbf{N}$, Theorems 2.2 and 2.3 allow one to build an (extended) non-recursive behavioural model in an arbitrary number of steps, without any reference to a recursive behavioural model. In particular, the same results allow the possibility of constructing an (extended) non-recursive behavioural model in an arbitrary number of steps, starting from an (extended) non-recursive behavioural model in a given number of steps, without any reference to a recursive behavioural model. Although a recursive behavioural model is not required for these constructions, it exists as shown by the proofs of the two theorems, as subsidiarity. ■

PART II.

STATE CONTROLLABILITY ANALYSIS

CHAPTER 3

TOWARDS THE EXPERIMENTAL ANALYSIS OF STATE CONTROLLABILITY

3.1. Theoretic results based on linear non-recursive behavioural models

Given a real plant (Σ), consider the continuous-time linear recursive behavioural model (1.4). Suppose that the minimal polynomial of matrix A in model (1.4) is of the degree $r \leq n$. In Voicu and Pastravanu (1987) the following result is derived, representing a background for the experimental analysis of the state controllability of a real plant (Σ).

Theorem 3.1

For almost every sampling and holding period $T > 0$, and for every linear non-recursive behavioural model (X_n, U_n) of a real plant (Σ), the following holds:

$$\operatorname{rank} X_r = \operatorname{rank}[B, AB, ..., A^{r-1}B], \tag{3.1}$$

Proof: According to the well-known result:

$$e^{At} = \sum_{k=1}^{r} f_k(t) A^{k-1},$$

where $f_k(t)$, $k = \overline{1,r}$, are determinable and linearly independent functions, relation (2.6), with (2.4) and $E = e^{AT}$, $H = \int_0^T e^{At} dtB$, may be rewritten as follows:

$$X_r = [B, AB, ..., A^{r-1}B][F(T) \otimes I_m]U_r, \tag{3.2}$$

where I_m is the m-th order unit matrix, $\otimes$ denotes the Kronecker product, and

$$F(T)=\begin{bmatrix} \int_0^T f_1(T-\theta)d\theta \ldots \int_0^T f_1(rT-\theta)d\theta \\ \vdots \qquad\qquad \vdots \\ \int_0^T f_r(T-\theta)d\theta \ldots \int_0^T f_r(rT-\theta)d\theta \end{bmatrix}.$$

In Troch (1971a) it is proved that $F(T)$ is non-singular for every $T>0$, satisfying:

$$\mathrm{Im}[\lambda_i(A)-\lambda_j(A)]\neq 2k\pi/T,\ k=0,\pm1,\pm2,..., \quad (3.3)$$

for

$$\mathrm{Re}[\lambda_i(A)-\lambda_j(A)]=0,$$

where $\lambda_i(A),\ i=\overline{1,n},$ are the eigenvalues of the matrix A. It follows that the matrix $F(T)\otimes I_m$ is non-singular for almost every sampling and holding period $T>0$, i.e., statement (3.1) holds for almost every $T>0$ and for any linear non-recursive behavioural model of a real plant (Σ). ■

The practical usefulness of Theorem 3.1 (when r is unknown, but n is known) is based on the following result.

Corollary 3.1

For almost every sampling and holding period $T>0$, and for every linear non-recursive behavioural model (X_n, U_n) of a real system (Σ), the following holds:

$$\mathrm{rank}\,X_n=\mathrm{rank}[B, AB, ..., A^{n-1}B].$$

Proof: It is analogous (*mutatis mutandis*) with the proof of Theorem 3.1. ■

This result represents the theoretic background of the experimental procedure and devices realized by Voicu et al. (1986 a, b) to allow the direct testing of the state controllability of a real plant (Σ) which in fact represents the starting point of defining and effectively using the linear non-recursive behavioural models.

3.2. Applications

a. Determination of the behavioural model order

It is initially possible that the order n of a real plant (Σ), is unknown. Its determination may rely on the results of section 3.1.

Assume that the behavioural model (1.4) is a minimal realization for a real plant (Σ). The choice of a state vector candidate $x \in \mathbf{R}^N$, with $N \in \mathbf{N}$, for which the inequality

$$\operatorname{rank} X_N < N,$$

holds true for every $T > 0$ (in practice, for many values $T > 0$) yields the conclusion that N is not the actual dimension of the minimal realization of a real plant (Σ). This is possible and plausible due to a "final" over-description of the plant (Σ), see Voicu and Pastravanu (1988), i.e., eventually, some state variables in $x \in \mathbf{R}^N$ possibly become redundant (the components of vector function $x(t)$ are not linearly independent).

As a matter of fact, in order to use the "trial and error" procedure, all the signals that may be considered as internal variables of the plant (Σ) are included in the state vector $x \in \mathbf{R}^N$ from the very beginning (i.e., from the commencement of data acquisition for the construction of a linear non-recursive behavioural model). Thus, by repeating the state controllability test discussed above and excluding (if necessary), step by step, the redundant variables (if any) from the vector function $x(t)$, one eventually finds a certain $x \in \mathbf{R}^n$ of size $n \leq N$, when the equality:

$$\operatorname{rank} X_n = n$$

is valid for almost every value of the sampling and holding period $T > 0$. This stage allows the establishment of the order of the minimal realization of the plant (Σ).

b. Determination of the structural indices

Consider equalities of the form (2.6), which express the connection between the linear recursive behavioural model and the non-recursive one. For every non-recursive behavioural model (X_n, U_n), with any orthogonal test signals and almost every sampling and holding period $T > 0$, the state controllability matrix R_n of the discrete-time linear behavioural model (1.5) is obtained with:

$$R_n = \alpha^{-2} X_n U_n^{\tau} . \tag{3.4}$$

This allows the determination of the controllability index $n_\delta(T)$ of the behavioural model (1.5).

Assume that the continuous-time behavioural model (1.4) has the controllability index μ. According to (Troch, 1971 b), starting from the origin and using piecewise, constant control inputs (i.e., controls of the form (2.1)) with the sampling period $T > 0$ (except for the values specified in (3.3)), the plant (1.4) may be controlled so as to reach any final state in $\mathbf{R}^n$, in the time μT. This result underlines the relationship between the controllability indices of the continuous-time and discrete-time behavioural models, i.e., $n_\delta(T) \leq \mu$. Obviously, from the point of view of the digital control of the real plant (Σ), it is of great interest to know $n_\delta(T)$ which corresponds to the used sampling and holding period T and especially:

$$\mu = \max_{T>0} n_\delta(T) . \tag{3.5}$$

The knowledge of R_n given by (3.4), also allows the determination of the structural indices and the characteristic parameters, see Ionescu (1986), of the discrete-time linear behavioural model (1.5) corresponding to the used T, without any information about matrices E and H in (1.5).

3.3. Numerical aspects of state controllability testing based on non-recursive behavioural models

The practical application of Theorem 3.1 and Corollary 3.1 relies on the singular values decomposition (SVD), which represents the most appropriate numerical procedure for rank computation according to Forsythe et al. (1977) and Golub and Van Loan (1983). Under these circumstances, it is necessary to state a rejection criterion for those non-zero singular values which are, in fact, insignificant for rank computation.

To simplify the writing of the following preparatory exposition, let us substitute X_n by $X \triangleq (x_{ij})$ and denote $\rho \triangleq nm$.

In Voicu and Pastravanu (1987) it has been shown that the SVD of matrix X may be influenced by two kinds of errors:

1. The measurement errors reflected in the elements of X, i.e., instead of matrix X one acquires a different matrix $\tilde{X} \triangleq (\tilde{x}_{ij})$.

2. The round-off errors due to floating-point arithmetic, which determine the computation of the singular values of a slightly modified matrix $\tilde{\tilde{X}}$.

The three matrices X, $\tilde{X}$ and $\tilde{\tilde{X}}$ are related by the following inequality:

$$\left\|X - \tilde{\tilde{X}}_2\right\| \leq \left\|X - \tilde{X}\right\|_2 + \left\|\tilde{X} - \tilde{\tilde{X}}_2\right\|.$$

Using the properties of the norms, one can evaluate upper bounds for both

$\|X-\tilde{X}\|_2$ and $\|\tilde{X}-\tilde{\tilde{X}}\|_2$ in the following manner:

$$\|X-\tilde{X}\|_2 \leq \sqrt{n\rho}\cdot\max|x_{ij}-\tilde{x}_{ij}| \leq \sqrt{n\rho\varepsilon}\cdot\max|\tilde{x}_{ij}| \leq$$
$$\leq \sqrt{n\rho\varepsilon}\|\tilde{X}\|_2 = \text{MEAC}\|\tilde{X}\|_2 .$$

ε represents the accuracy of the measuring instrumentation and $\text{MEAC} = \sqrt{n\rho\varepsilon}$.

On the other hand, one can only evaluate the singular values of $\tilde{\tilde{X}}$. But according to Dongarra et al. (1977) one can write:

$$\|\tilde{\tilde{X}}-\tilde{X}\|_2 \leq \text{COAC}\|\tilde{X}\|_2 ,$$

where COAC is the product of the machine epsilon and a moderately increasing function of variables n, ρ.

So, one can further write:

$$\|\tilde{X}\|_2 \leq \|\tilde{X}-\tilde{\tilde{X}}_2\| + \|\tilde{\tilde{X}}\|_2 \leq \text{COAC}\|\tilde{X}\|_2 + \|\tilde{\tilde{X}}\|_2$$

and the first inequality of this approach becomes:

$$\|X-\tilde{\tilde{X}}\|_2 \leq \frac{\text{MEAC+COAC}}{\text{1-COAC}}\|\tilde{\tilde{X}}\|_2 .$$

An important property of singular values is their rather low sensitivity with respect to changes in the elements of the corresponding matrix. Let us denote by:

$$s_1 \geq s_2 \geq \ldots \geq s_n \geq 0,\ s_i > 0,$$

the singular values of the matrix X and by:

$$\tilde{\tilde{s}}_1 \geq \tilde{\tilde{s}}_2 \geq \ldots \geq \tilde{\tilde{s}}_n \geq 0,\ \tilde{\tilde{s}}_1 > 0,$$

the computed ones, i.e., the singular values of the matrix $\tilde{\tilde{X}}$.

According to Klema and Laub (1986), the SVD algorithm is quite stable and the following evaluation holds:

$$\left|s_i - \tilde{\tilde{s}}_i\right| \leq \left\| X - \tilde{\tilde{X}} \right\|_2 \leq \mathrm{AC}\tilde{\tilde{s}}_1,\ i = \overline{1,n},$$

where:

$$\mathrm{AC} = \frac{\mathrm{MEAC}+\mathrm{COAC}}{1-\mathrm{COAC}}. \tag{3.6}$$

Using now the initial notation X_n from the linear non-recursive behavioural model, one may conclude that:

1. If

   $$\tilde{\tilde{s}}_n / \tilde{\tilde{s}}_1 > \mathrm{AC},$$

   then $\operatorname{rank} X_n = n$, i.e., the system is completely state-controllable.

2. If there exists $k < n$ such that:

   $$\tilde{\tilde{s}}_k / \tilde{\tilde{s}}_1 > \mathrm{AC}$$

   and

   $$\tilde{\tilde{s}}_{k+1} / \tilde{\tilde{s}}_1 \leq \mathrm{AC},$$

   then $\operatorname{rank} X_n = k$, and one has to eliminate the eventually redundant state variables from the state vector candidate previously considered (see also subsection 3.2.a).

It is obvious that in (3.6), for the usual data acquisition and processing equipment, one may consider MEAC $\gg$ COAC and COAC $\ll 1$, i.e., from (3.6) it results that AC $\cong$ MEAC.

Note that for any choice of orthogonal signals $U_n = \alpha\Theta$, the value of the ratio s_n / s_1 does not change. This value characterizes the numerical conditioning of the controllability matrix R_n. In this respect one can prove the following result.

Theorem 3.2

For any linear non-recursive behavioural model (X_n, U_n) of the sampling period $T > 0$, associated with the real plant (Σ), in the case of orthogonal test signals $U_n = \alpha\Theta$, the following equality holds:

$$\frac{s_i}{s_1} = \frac{r_i}{r_1},\ i = \overline{1,n},$$

where $s_1 \geq s_2 \geq ... \geq s_n \geq 0, s_1 > 0,$, and $r_1 \geq r_2 \geq ... \geq r_n \geq 0, r_1 > 0,$ denote the singular values of matrices X_n and R_n respectively.

Proof: This results from the fact that for any orthogonal $\Theta \in \mathbf{R}^{nm \times nm}$ and $\alpha > 0$ one can write:

$$s_i = \alpha r_i,\ i = \overline{1,n}. \qquad \blacksquare$$

Remark 3.1

The ratios $r_i / r_1,\ i = \overline{1,n},$ depend on the sampling and holding period $T > 0$. From a practical point of view, it is significant that there exist values of T which maximize the minimal singular value r_n of R_n. Indeed, according to Theorem 2.1, for $\alpha = 1$, the linear independency of the columns of the matrix $R_n = X_n$ is "precarious" both for small and large values of T (when nT is comparable with the stabilization time of the step response). It obviously results that:

$$\lim_{T\to 0} r_n(T) = 0, \quad \lim_{T\to\infty} r_n(T) = 0$$

and as $r_n(T)$ is a continuous and positive function, it follows that $r_n(T)$ has at least a local maximum for $T \in (0, +\infty)$. If the linear recursive behavioural model (1.4) has complex eigenvalues, then at each interval (T_1, T_2), $r_n(T)$ has at least a local maximum, where $0 < T_1 < T_2$ are two consecutive values of T for which matrix $F(T)$ is singular (see the proof of Theorem 3.1).

Similar considerations may be formulated regarding the maximization of the ratio $r_n(T)/\ r_1(T)$. Practically, one will acquire data for many values of T (nT being less than the step response stabilization time) and in this way one may choose the value of T to maximize $r_n(T)$ or $r_n(T)/r_1(T)$. ■

Remark 3.2

Theorem 3.2 provides evidence that the orthogonal test signals defined in section 2.2 do not influence the state controllability investigation because ratio r_n / r_1, characterizes the plant behaviour and not the evolution of the experiments for controllability testing. Ratio r_n / r_1, is strictly connected to the geometric mechanism of the state controllability and this is also the reason for evaluating the rank of R_n (when the discrete-time linear recursive behavioural model (1.5) is known) using SVD because r_n / r_1, represents the numerical conditioning of R_n. More details on this subject will be given in the next section. ■

Remark 3.3

Orthogonal test signals allow a numerically advantageous determination of the state controllability matrix R_n, according to (3.4), because $U_n^{-1} = = \alpha^{-1}\Theta^{\tau}$. ■

Remark 3.4

Under these circumstances, the state controllability of any real plant (Σ) may be directly (i.e., experimentally) tested as in Voicu et al. (1986 a) and Voicu et al. (1986 b) and this fact represents a consistent physical and experimental basis for the notion of state controllability according to Voicu and Pastravanu (1987). ■

SVD also allows the determination of the structural indices by testing the linear independency of the columns of the state controllability matrix R_n.

3.4. Evaluation of the quality of state controllability via energy criteria

Since Kalman introduced the notion of state controllability, some "figures of merit" to express the quality of state controllability have been defined by Kalman et al. (1963), Johnson (1969), Kreindler and Saracik (1964) and Moore (1981). In this way, it is possible to compare two systems or a system with itself when its parameters are modified under some environmental conditions.

The significance of the ratio s_n / s_1 associated with any linear non-recursive behavioural model (X_n, U_n), with orthogonal test signals, is evidenced in Voicu and Pastravanu (1987) via the energy functional, allowing a pertinent evaluation of state controllability quality.

In this respect, for the continuous-time linear system (S) described by the linear and recursive mathematical behavioural model (1.4), assumed to be completely state-controllable, the following two forms of state controllability degrees can be defined, for a finite and fixed time horizon $t_f > 0$:

$$1^{\circ}\ \mathrm{cod}_1 S(t_f) = \frac{\min_{\|x(t_f)\|_2=1} \int_0^{t_f} \left\| u_{ox_f}(t) \right\|_2^2 dt}{\max_{\|x(t_f)\|_2=1} \int_0^{t_f} \left\| u_{ox_f}(t) \right\|_2^2 dt} \tag{3.7}$$

where $u_{ox_f}(t)$ represents the optimal control, for the initial condition $x(0) = 0$ and the final condition $x_f = x(t_f) \neq 0$, with respect to the energy functional:

$$J = (1/2)\int_0^{t_f} \|u(t)\|_2^2 dt, \tag{3.8}$$

and

$$2^{\circ}\ \mathrm{cod}_2 S(t_f) = \frac{\min_{J=1} \left\| x(t_f) \right\|_2}{\max_{J=1} \left\| x(t_f) \right\|_2}. \tag{3.9}$$

For piecewise constant inputs, with a sampling period $T > 0$, and for $t_f = nT$, the evaluation of state controllability degrees $\mathrm{cod}_1 S(t_f)$ and $\mathrm{cod}_2 S(t_f)$ is approached by the following theorem.

Theorem 3.3

For any $t_f = nT$, with $T > 0$ according to Theorem 3.1, the following equalities hold:

$$\mathrm{cod}_1 S(nT) = (r_n / r_1)^2, \tag{3.10}$$

$$\mathrm{cod}_2 S(nT) = r_n / r_1, \tag{3.11}$$

where $r_1 \geq r_2 \geq \ldots \geq r_n > 0$ denote the singular values of the state controllability matrix R_n associated with the discrete-time linear recursive behavioural model (1.5).

Proof: For any final state $x_n = x(nT) \in \mathbf{R}^n$ there exists an optimal control u_{ox_n}, piecewise constant, with a period $T > 0$, which minimizes the energy functional (3.8):

$$J = \frac{1}{2}\int_0^{nT} \|u(t)\|_2^2 dt = \frac{1}{2}\sum_{i=1}^{n} \int_{(i-1)T}^{iT} \|u_i\|_2^2 dt = \frac{T}{2}\|\overline{u}_n\|_2^2. \quad (3.12)$$

Therefore, minimization of the energy functional J is equivalent to the minimization of $\|\overline{u}_n\|_2^2$. Under these circumstances, using (2.5) for $j = n$, one obtains the optimal control:

$$\overline{u}_{ox_n} \triangleq \overline{u}_n = R_n^+ x_n = R_n^\tau (R_n R_n^\tau)^{-1} x_n,$$

where R_n^+ is the Moore-Penrose pseudoinverse of R_n. Thus, one may write:

$$\begin{aligned} J &= \frac{T}{2}\|\overline{u}_n\|_2^2 = \\ &= x_n^\tau [R_n^\tau (R_n R_n^\tau)^{-1}]^\tau [R_n^\tau (R_n R_n^\tau)^{-1}] x_n = x_n^\tau (R_n R_n^\tau)^{-1} x_n. \end{aligned} \quad (3.13)$$

Now, from (3.7), with $t_f = nT$ and (3.13), the result is:

$$\operatorname{cod}_1 S(nT) = \frac{\min_{\|x_n\|_2=1} x_n^\tau (R_n R_n^\tau)^{-1} x_n}{\max_{\|x_n\|_2=1} x_n^\tau (R_n R_n^\tau)^{-1} x_n} = (r_n / r_1)^2,$$

i.e., (3.10) holds.

To prove (3.11), taking into account $J = 1$ with (3.12) yields:

$$\|\overline{u}_n\|_2^2 = 2 / T,$$

and from (3.9), with $t_f = nT$, and (2.5) for $j = n,$ one obtains:

$$\mathrm{cod}_2 S(nT) = \frac{\min_{\|\overline{u}_n\|_2^2=2/T} \|R_n \overline{u}_n\|_2}{\max_{\|\overline{u}_n\|_2^2=2/T} \|R_n \overline{u}_n\|_2} = \frac{r_n}{r_1},$$

i.e., (3.11) holds. ■

Using Theorem 3.2, one can easily formulate a result similar to Theorem 3.3, but applicable to the linear non-recursive behavioural model of the real system (Σ).

Theorem 3.4

For any linear non-recursive behavioural model (X_n, U_n) of a real plant (Σ), with $T > 0$ according to Theorem 3.1 and with orthogonal test signals $U_n = \alpha\Theta,$ the following equalities hold:

$$\mathrm{cod}_1 S(nT) = (s_n / s_1)^2, \tag{3.14}$$

$$\mathrm{cod}_2 S(nT) = s_n / s_1, \tag{3.15}$$

where $s_1 \geq s_2 \geq ... \geq s_n > 0$ denote the singular values of the matrix X_n. ■

Remark 3.5

Theorems 3.3 and 3.4 allow a quantitative evaluation of the state controllability quality of a completely state controllable system for a given finite and fixed time horizon t_f. Clearly, the state controllability degrees (3.7) and (3.9) depend on t_f. Moreover, Theorem 3.4 opens the possibility of experimentally analysing the state controllability quality of a real plant (Σ).

The state controllability degree $\mathrm{cod}_1 S / \mathrm{cod}_1 \Sigma$ makes it possible to observe any anisotropic dynamics of the analysed system in its state space $\mathbf{R}^n$. This means that, starting from the origin, the system can reach any state situated on a sphere in $\mathbf{R}^n$ (and centred on the origin) for a finite and fixed time horizon, but under a different optimal consumption of control energy, which obviously depends on the position of the final state on the mentioned sphere.

The same anisotropic dynamics may be highlighted by the state controllability degree $\mathrm{cod}_2 S / \mathrm{cod}_2 \Sigma$ but this time the available control energy is fixed and the same for any state transfer from the origin to the final state in $\mathbf{R}^n$. Correspondingly, the reachable final states, using the same control energy each time, are situated at different distances with respect to the origin and these distances depend on the direction of the final state vector. ■

From (3.10), (3.14) and (3.11), (3.15) it follows that the numerical conditioning of R_n and X_n respectively, is not only a simple algebraic fact or symptom reflecting the state controllability quality, but also evidences the intimate mechanism and makes very transparent the geometric background of the quality of state controllability.

The fact that all the above results may be easily obtained using linear non-recursive behavioural models with orthogonal test signals is another reason for the naturalness of such an approach.

3.5. Non-recursive behavioural models and Hankel block matrices

Consider a completely state controllable system of the order n, sampled with a period that satisfies (3.3). For any non-recursive behavioural model in n steps, (X_n, U_n), the equality:

$$\mathrm{rank}\, X_n = n \tag{3.16}$$

holds and makes it possible to further explore the characterization conditions given in Theorems 2.2 and 2.3, for non-recursive behavioural models in an arbitrary number of steps $j > n$. In the remainder of this section the same notations as in section 2.5 will be used.

Corollary 3.2

The pair of matrices (X_j, U_j), with arbitrary $j > n$, is a non-recursive behavioural model in j steps if and only if:

$$\operatorname{rank}\begin{bmatrix} \gamma_1 & \cdots & \gamma_{j-1} \\ \gamma_2 & \cdots & \gamma_j \end{bmatrix} = n, \ \forall j \geq n+1. \tag{3.17}$$

Proof: This follows from Theorem 2.2, taking into account that (3.16) involves the rank condition:

$$\operatorname{rank}[\gamma_1, ..., \gamma_n] = n. \qquad \blacksquare$$

Corollary 3.3

The set of matrices $(X_{j+l}, ..., X_j, U_j)$, with arbitrary $j > n$, is an extended non-recursive behavioural model in $j+l$ steps if and only if:

$$\operatorname{rank}\begin{bmatrix} \gamma_1, ..., \gamma_{j-1}, X_j ..., X_{j+l-1} \\ \gamma_2, ..., \gamma_j, X_{j+1} ..., X_{j+l} \end{bmatrix} = n, \ \forall j \geq n+1, \ \forall l \in \mathbf{N}. \tag{3.18}$$

Proof: This follows from Theorem 2.3, taking into account that (3.16) involves the rank condition:

$$\operatorname{rank}[\gamma_1, ..., \gamma_{j-1}, X_j ..., X_{j+l-1}] = n. \qquad \blacksquare$$

Remark 3.6

Corollary 3.2 gives a necessary and sufficient condition for the existence and unicity of a recursive behavioural model (1.5) such that the pair of matrices (X_j, U_j), $j > n$, represents a non-recursive behavioural model built from (1.5) in accordance with Definition 2.1. Similarly, Corollary 3.3 gives a necessary and sufficient condition for the existence and unicity of a recursive behavioural model (1.5) such that the set of matrices $(X_{j+l},\ldots$ $\ldots,X_j,U_j)$, $j>n$, represents an extended non-recursive behavioural model built from (1.5) in accordance with Definition 2.3. ■

Since we have considered all the state variables accessible to measurement, i.e., the output vector coincides with the state vector, $C=I_n$ in (1.5). Therefore matrices $\gamma_i \in \mathbf{R}^{n\times m}$, $i=\overline{1,j}$, are the first j Markov parameters of' the transfer matrix function of (1.5). Denoting generically by $H_{p,q}$ the Hankel block matrix:

$$H_{p,q}=\begin{bmatrix}\gamma_1,\ldots\ldots,\gamma_q \\ \ldots\ldots \\ \gamma_p,\ldots\ldots\gamma_{p+q-1}\end{bmatrix}, \tag{3.19}$$

according to Kalman et al. (1969), the equalities:

$$\text{rank}\, H_{n+\nu,n+\mu}=n,\ \forall\nu\geq 0,\ \forall\mu\geq 0, \tag{3.20}$$

hold because the system is completely state controllable.

The following algebraic result will be able to reveal the equivalence between the properties of the non-recursive behavioural models and the properties of Hankel block matrices.

Proposition 3.1

If $\gamma_i \in \mathbf{R}^{n\times m}$, $i\in\mathbf{N}$, is a sequence of matrices satisfying:

$$\text{rank}[\gamma_1,\ldots,\gamma_n]=n, \tag{3.21}$$

Then the rank condition (3.20) is equivalent to

$$\text{rank} H_{2,n+\mu}=n,\ \forall\mu\geq 0. \tag{3.22}$$

Proof: If (3.20) is true, then $\text{rank}\, H_{2,n+\mu} \leq n$, since $H_{2,n+\mu}$ is a block of $H_{n,n+\mu}$. Taking into account (3.21) results in $\text{rank} H_{2,n+\mu} = n$, i.e., (3.22) is proved.

Conversely, if (3.22) is true, the addition of new rows of blocks to $H_{2,n+\mu}$ does not change the rank of this matrix, because all new row vectors are linearly dependent on the row vectors of the matrix $[\gamma_1, ..., \gamma_n]$. Thus (3.20) is fulfilled. ■

Remark 3.7

Using the properties of Hankel block matrices, Proposition 3.1 supplies the same result as derived in Corollary 3.2, by operating with the conditions that characterize non-recursive models. ■

Finally, we have to point out that the approach based on Markov parameters and Hankel block matrices, despite its equivalence to non-recursive behavioural models, is able to create a broader perspective for determining the model order, which in this section has been considered as *a priori* known. This fact will be exploited in the next section.

3.6. Non-recursive behavioural models based on finite sequences of experimental data

In practice, the real order of a system may be unknown initially and section 3.2.a has presented a procedure for model order determination, based on the state controllability. As already mentioned, this procedure operates in a "trial and error" manner and therefore any supplementary aid in finding a starting value might be precious. In this respect, the connection between the non-recursive behavioural models and the Hankel block matrices built from the Markov parameters may be further explored under the following assumption: there exists a non-recursive behavioural model in j steps (j arbitrary), whose real order n is unknown, and instead of the matrix

$X_j \in \mathbf{R}^{n\times mj}$ (see Definition 2.1) the experiments provide a matrix $Y_j \in \mathbf{R}^{p\in mj}$. Here the matrix Y_j plays the role of a candidate for the state space vector and can be regarded as the output vector of the discrete-time behavioural model (1.5) whose real state vector is not known. Similar to (2.6) one can write:

$$Y_j = G_j U_j, \tag{3.23}$$

where $G_j \in R^{p\times nj}$ denotes the output controllability matrix. G_j may be partitioned in the same way as Γ_j in (2.23), yielding:

$$G_j = [g_1, ..., g_j],\ g_j \in \mathbf{R}^{p\times m},\ i = \overline{1, j}. \tag{3.24}$$

Now if one constructs the Hankel block matrix $H_{p,q}$,

$$H_{p,q} = \begin{bmatrix} g_1, \ldots\ldots, g_q \\ \cdots\cdots \\ g_p, \ldots\ldots, g_{p+q-1} \end{bmatrix}, \tag{3.25}$$

there exists an integer n such that the rank condition (3.20) will be fulfilled. According to Kalman et al. (1969), this integer value n represents the actual order of the system, which was initially unknown. Obviously, the determined n depends on the choice of the candidate vector, because the resulting discrete-time system (1.5), with the candidate vector as the output vector, may be just incompletely observable; this aspect pertains to the ability of the experimenter. In general, for a well-designed experiment, this value of n can stand for the starting point in the determination of the model order (using, for instance, the procedure in section 3.2.a). Moreover, this initial value considered for n can give a hint about the fulfilment of the working hypothesis that all the state variables are accessible to measurement.

At this point it is necessary to draw attention to another practical problem the experimenter must cope with, namely the comprehensive exploitation of a finite sequence of experimental data. As discussed before, the usage of the Markov parameters for getting an estimation of the model order requires (3.20) to be true. Since just finite sequences of experimental data are available, a finite number of rank conditions of the form (3.20) can be tested. In these realistic conditions the only way to approach the model order determination is the realizability criterion in Kalman et al. (1969), which ensures the existence of a partial realization corresponding to a finite number of terms from the Markov parameter sequence. According to Kalman et al. (1969), this partial realization corresponds to a rational approximation which serves as a linear behavioural model of the real system behaviour. Thus, a finite value of the model order can always be found. It is worth underlining that such a point of view is fully compatible with properties of the non-recursive behavioural models which have been discussed previously. Hence a non-recursive behavioural model in j steps resulting from a finite set of experimental data can be expanded to an arbitrary number of steps using Theorem 2.2 or Corollary 3.2. In other words, for any real system a non-recursive behavioural model in an arbitrary number of steps does exist, but the experimenter should decide whether its accuracy is acceptable for the envisaged purpose. It should also be mentioned that although the properties of the non-recursive behavioural models are strongly related to the Markov parameters, the scenario constructed in this work relies on their exploitation as a global entity. Unlike other approaches which are using the Markov parameters for building a (more or less accurate) recursive behavioural model, our emphasis is exclusively placed on devising analysis and synthesis algorithms in terms of non-recursive behavioural models.

A final remark refers to the numerical computation of the rank conditions (required by different theoretical results), which should follow (*mutatis mutandis*) the strategy presented in section 3.3.

CHAPTER 4

EFFECTS OF PIECEWISE CONSTANT AND QUANTIZED CONTROL ON THE GEOMETRY OF THE REACHABILITY SET

Usually, digital control with a finite and fixed time horizon $t_f > 0$ may be relaxed with respect to the restriction $t_f = nT$ which has been stated by the evaluation of the state controllability quality in section 3.4. For $T>0$ (according to Theorem 3.1), consider $t_f = NT$, $N \in \mathbf{N}$, $N \geq \mu$, where μ denotes the controllability index – see subsection 3.2 of the continuous-time linear recursive behavioural model (1.4) and at the same time the maximum value of the controllability index of the discrete-time linear recursive behavioural model (1.5); and note that μ can be experimentally determined (see subsection 3.2.b and (3.5)). The geometry of the reachability set for a finite and fixed $t_f > 0$ and for energy or amplitude type constraints depends on the sampling and holding period T as proven in Pastravanu and Voicu (1992c). The linear non-recursive behavioural model, with orthogonal test signals, allows the analysis of the geometry of the reachability set using only data directly acquired by measurements.

4.1. Case of energy constrained control

Let us consider the completely state controllable plant (Σ) described by the continuous-time linear recursive behavioural model (1.4) and controlled by input signals satisfying the energy type constraint:

$$\int_0^{t_f} \|u(t)\|_2^2 dt \leq 1 \tag{4.1}$$

for the finite and fixed time horizon $t_f > 0$.

For piecewise constant input signals, with a sampling and holding period $T > 0$, the geometry of the reachability set which corresponds to the finite and fixed $t_f > 0$ and to the constraint (4.1) is characterized by the following two theorems.

Theorem 4.1

For any finite and fixed time horizon $t_f > 0$ and for any sampling and holding period $T > 0$ (according to Theorem 3.1) satisfying $t_f = NT$, $N \geq \mu$, (μ denoting the controllability index – see subsection 3.2.b) and under the constraint (4.1), the reachability set of the plant (Σ), described by the linear recursive behavioural model (1.4) and, controlled by piecewise constant inputs with a sampling and holding period T, is a hyper-ellipsoid $(E) \in \mathbf{R}^n$ with the semi-axes given by:

$$a_i = \frac{r_i}{\sqrt{T}} v_i, \; i = \overline{1,n}. \tag{4.2}$$

Vectors v_i, $i = \overline{1,n}$, are the columns of the matrix $V \in \mathbf{R}^{n \times n}$ of the SVD of matrix R_N (i.e., $j = N$ in (2.4)):

$$R_N = V \operatorname{diag}\{r_1, \ldots, r_n\} W^{\tau} \in \mathbf{R}^{n \times Nm}, \tag{4.3}$$

and $r_1 \geq r_2 \geq \ldots \geq r_n > 0$ are the singular values of R_N.

Proof: Constraint (4.1) may be rewritten as:

$$\sum_{i=1}^{N} \int_{(i-1)T}^{iT} \|u_i\|_2^2 dt \leq 1$$

or equivalently (with (2.3) and $j = N$) :

$$\sqrt{T}\left\|\overline{u}_N\right\|_2 \leq 1. \qquad (4.4)$$

According to (2.5), for $j = N$, the transfer of the state from the origin to x_N is described by:

$$x_N = R_N \overline{u}_N \qquad (4.5)$$

and the convex set defined by (4.4) is transformed by (4.5) into another convex set, $(E) \in \mathbf{R}^n$, which represents the reachability set of the system. According to Golub and Van Loan (1983), (E) is a hyper-ellipsoid with the semi-axes given by (4.2). As R_N represents a surjective function, it results that for any $x_N \in (E)$ there exists a piecewise control with the sampling and holding period T, satisfying constraint (4.1), which ensures the state transfer from the origin to x_N within the total time $t_f = NT$. ■

The effects of decreasing the sampling and holding period T are outlined by the following result.

Theorem 4.2

Denote by (E) and (E') two reachability hyper-ellipsoids corresponding to the sampling and holding periods $t_f = NT$, $N \geq \mu$, and $T' = t_f/\nu N$, $\nu \in \mathbf{N}$, $\nu \geq 2$, respectively.

Then:

$$(E) \subseteq (E') \qquad (4.6)$$

and for each pair of homologous semi-axes the inequality

$$\left\|a_i\right\|_2 \le \left\|a_i'\right\|_2, \ i=\overline{1,n}, \tag{4.7}$$

holds.

Proof: Let $u_0, u_1, \dots, u_{N-1}$ be any control input sequence with the sampling and holding period T, satisfying constraint (4.1), or equivalently let us consider $\overline{u}_N \in \mathbf{R}^{Nm}$ satisfying constraint (4.4).

This control sequence transfers the state of the system from the origin to $x^* \in (E)$.

But the same control sequence may be seen as a control sequence with the sampling and holding period T', with $\overline{u}_{\nu N} \in \mathbf{R}^{\nu Nm}$, $u_{(k-1)\nu+1} = \dots = u_{k\nu}$, $k = \overline{1,N}$, which satisfies (according to (4.1)):

$$\sqrt{T/\nu}\left\|\overline{u}_{\nu N}\right\|_2 \le 1$$

and transfers the system state from the origin to $x^* \in (E)$.

In other words, any point of (E) also belongs to (E'), i.e., (4.6) is proved.

The norms of the semi-axes of (E) are given (according to 4.2) by:

$$\left\|a_i\right\|_2 = \sqrt{1/T}\, r_i, \ i=\overline{1,n}. \tag{4.8}$$

Similarly (and *mutatis mutandis*) for (E') one can write:

$$\left\|a_i'\right\|_2 = \sqrt{\nu/T}\, r_i', \ i=\overline{1,n}, \tag{4.9}$$

where r_i', $i=\overline{1,n}$, are the singular values of $R_{\nu N}$.

According to the result proved in Appendix B, between R_N and $R_{\nu N}$ there exists the following relation:

$$R_N = R_{\nu N} M_{\nu N},$$

where

$$M_{\nu N} = \mathrm{diag}\{\underbrace{M_I, \ldots, M_I}_{N \text{ blocks}}\} \in \mathbf{R}^{\nu Nm \times Nm},$$

and

$$M_I^{\tau} = [\underbrace{I_m, \ldots, I_m}_{\nu \text{ blocks}}] \in \mathbf{R}^{m \times \upsilon m}$$

I_m is the unit matrix of order m.

Moreover, one can write:

$$\begin{bmatrix} R_N & 0 \\ 0 & 0 \end{bmatrix} = \begin{bmatrix} R_{\nu N} \\ 0 \end{bmatrix} \begin{bmatrix} M_{\nu N} & 0 \end{bmatrix},$$

where the three block matrices are square matrices of appropriate dimensions. Now using the result proved in Appendix C, referring to the singular values of the product of two square matrices, one obtains:

$$r_i \le \sqrt{\nu}\, r_i', \; i = \overline{1, n},$$

This is because the first n singular values of $\begin{bmatrix} R_N & 0 \\ 0 & 0 \end{bmatrix}$ and $\begin{bmatrix} R_{\nu N} \\ 0 \end{bmatrix}$ are $r_1 \ge r_2 \ge \ldots \ge r_n$ and $r_1' \ge r_2' \ge \ldots \ge r_n'$ respectively, whereas all non-zero singular values of $[M_{\nu N} \;\; 0]$ are equal to $\sqrt{\nu}$. Thus, by considering equalities (4.8) and (4.9), we get inequality (4.7). ■

This result completes that one in Moore (1981) regarding the influence of the sampling and holding period on the geometry of the reachability set of the system (1.4) for a finite and fixed time horizon t_f and under constraint (4.1). From Proposition 7 presented in Moore (1981) it easily follows that decreasing the sampling period T, for T approaching zero, the sequence of reachability hyper-ellipsoids converges to a hyper-ellipsoid whose semi-axes have norm 2 equal to the singular values of the state controllability Gramian. This fact yields the following corollary.

Corollary 4.1

For any control input constrained by (4.1) and for any finite and fixed time horizon $t_f > 0$, the reachability set of the completely state controllable system (1.4) is a hyper-ellipsoid whose semi-axes are given by the SVD of the Gram matrix:

$$\int_0^{t_f} e^{At} BB^{\tau} e^{A^{\tau} t} dt.$$

This includes all the reachability sets corresponding to a piecewise constant control (2.1), for almost every sampling and holding period $T>0$ (according to Theorem 3.1). For T approaching zero, it represents the limit of the corresponding sequence of the reachability sets. ■

The usefulness of the linear non-recursive behavioural models in analysing the geometry of the reachability set is shown by the following result.

Theorem 4.3

Let (E) be the reachability set of the completely state controllable linear recursive behavioural model (1.4), associated with the real plant (Σ), controlled by (2.1) and constrained by (4.1), with the sampling and holding period $T=t_f/N$, $N \geq \mu$, (according to Theorem 3.1) for a finite and fixed time horizon $t_f > 0$.

Let (X_N, U_N) be a linear non-recursive behavioural model with orthogonal test signals $U_N = \alpha\Theta$ associated with the real plant (Σ), with the sampling and holding period T.

Then $(E)$is a hyper-ellipsoid with the semi-axes given by:

$$a_i = (s_i / \alpha\sqrt{T})\bar{v}_i, \; i = \overline{1,n},$$

where $\bar{v}_i$, $i = \overline{1,n}$, are the columns of the matrix if $\bar{V} \in \mathbf{R}^{n \times n}$ of the SVD of the matrix X_N, i.e.,

$$X_N = \bar{V}\,\mathrm{diag}\{s_1,...,s_n\}\bar{W}^{\tau},$$

and $s_1 \geq s_2 \geq ... \geq s_n > 0$ denote the singular values of X_N.

Proof: R_N given by (4.3) is substituted in (2.6) for $j = N$ and (2.9) is used for expressing U_N. It follows that:

$$X_N = V\,\mathrm{diag}\{r_1,...,r_n\}W^{\tau}\alpha\Theta,$$

i.e., the SVD of X_N has the form:

$$X_N = V\,\mathrm{diag}\{\alpha r_1,...,\alpha r_n\}(\Theta^{\tau}W)^{\tau}.$$

The proof is completed using (4.1) and Theorem 4.1. ■

Finally, the relation between the semi-axes of the reachability hyper-ellipsoid and the state controllability degrees $\mathrm{cod}_1 S(t_f)$ and $\mathrm{cod}_2 S(t_f)$ (defined and characterized in section 3.4) can be as follows.

Theorem 4.4

Denote by a_i, $i = \overline{1,n}$, the semi-axes of the reachability hyper-ellipsoid according to Theorem 4.1, with $\|a_1\|_2 \geq \|a_2\|_2 \geq .. ... \geq \|a_n\| > 0$. Then there exist the equalities:

$$\operatorname{cod}_1 S(NT) = \frac{\|a_n\|_2^2}{\|\ \ _1\|_2^2}, \qquad \operatorname{cod}_2 S(NT) = \frac{\|a_n\|_2}{\|\ \ _1\|_2}.$$

Proof: For any final state $x_N = x(NT) \in \mathbf{R}^n$ there exists a piecewise constant optimal control with a period $T > 0$, denoted u_{oxT} which minimizes the energy functional:

$$J = (1/2)\int_0^{NT} \|u(t)\|_2^2 dt = (1/2)\sum_{i=1}^{N} \int_{(i-1)T}^{iT} \|u_i\|_2^2 dt = (T/2)\|\overline{u}_N\|_2^2.$$

Thus the minimization of J is equivalent to the minimization of $\|\overline{u}_N\|_2^2$ and from (3.7) with $t_f = NT$ and from (2.5) written for $j = N$, it follows that:

$$\operatorname{cod}_1 S(NT) = \frac{\min_{\|x_N\|_2=1} x_N^\tau (R_N R_N^\tau)^{-1} x_N}{\max_{\|x_N\|_2=1} x_N^\tau (R_N R_N^\tau)^{-1} x_N} = (r_n / r_1)^2,$$

where $r_1 \geq r_2 \geq \ldots r_n > 0$ are the singular values of the matrix $R_N \in \mathbf{R}^{n \times Nm}$. Now using Theorem 4.1 one obtains the first equality of the theorem.

In order to prove the second equality, one must take into account (4.8) and the constraint $J=1$. As it has been shown by the proof of Theorem 3.3, $J=1$ is equivalent (*mutatis mutandis*) to:

$$\|\overline{u}_N\|_2^2 = 2/T,$$

and from (3.9) with $t_f = NT$, and from (2.5) written for $j = N$ it results in:

$$\operatorname{cod}_2 S(NT) = \frac{\min_{\|\overline{u}_N\|_2^2=(2/T)} \|R_N \overline{u}_N\|_2}{\max_{\|\overline{u}_N\|_2^2=(2/T)} \|R_N \overline{u}_N\|_2} = r_n / r_1.$$

Now using Theorem 4.1 one obtains the second equality of the theorem. ■

An immediate consequence concerns $\mathrm{cod}_1 S$ and $\mathrm{cod}_2 S$ computed for any piecewise continuous control inputs.

Corollary 4.2

Let α_i, $i=\overline{1,n}$, be the semi-axes of the reachability hyper-ellipsoid according to Corollary 4.1, with $\|\alpha_1\|_2 \geq \|\alpha_2\|_2 \geq \ldots \geq \|\alpha_n\|_2 > 0$.

Then there exist the following equalities:

$$\mathrm{cod}_1 S(t_f) = \frac{\|\alpha_n\|_2^2}{\|\alpha_1\|_2^2}, \qquad \mathrm{cod}_2 S(t_f) = \frac{\|\alpha_n\|_2}{\|\alpha_1\|_2}.$$

Proof: This results from Theorem 4.2, cconsidering Corollary 4.1. ■

Remark 4.1

Using linear non-recursive behavioural models with orthogonal test signals and a sufficiently small sampling and holding period $T > 0$ (according to Theorem 3.1), one can determine approximations of the controllability degrees defined by (3.7) and (3.9), using either Theorem 3.4 or Theorem 4.4. ■

4.2. Case of amplitude constrained control

Let us consider the completely state controllable plant (Σ), described by the linear recursive behavioural model (1.4) and controlled by the amplitude constrained input signal:

$$\|u(t)\|_\infty \leq 1, \; t \in \mathbf{R}_+. \tag{4.10}$$

For piecewise constant control input (with the sampling and holding period $T > 0$) which satisfies (4.10), the geometry of the reachability set corresponding to a given finite time horizon $t_f > 0$ is characterized by the following two theorems.

Theorem 4.5

Consider a plant (Σ), described by the linear recursive behavioural model (1.4) and controlled by piecewise constant input signals with the sampling and holding period T (according to Theorem 3.1), constrained by inequality (4.10). Given a finite and fixed time horizon of $t_f > 0$, satisfying $t_f = NT$, $N \geq \mu$, the corresponding reachability set is a convex hyper-polyhedron $(P) \in \mathbf{R}^n$, that is symmetrical with respect to the origin, whose vertices are:

$$\sum_{k=1}^{N} \sum_{i=1}^{m} \sigma_{ki} E^{k-1} h_i \tag{4.11}$$

with suitable $\sigma_{ki} \in \{-1.1\}$ ensuring the convexity of (P) and h_i, $i=\overline{1,m}$, denoting the columns of matrix H in (1.5).

Proof: Taking into account the form of control (2.1), constraint (4.10) can be equivalently rewritten as:

$$\left\| \overline{u}_N \right\|_\infty \leq 1. \tag{4.12}$$

According to (2.5), for $j = T$, the result is:

$$x_N = R_N \overline{u}_N \tag{4.13}$$

and the convex set defined by (4.12) is transformed by (4.13) into another convex set $(P) \in \mathbf{R}^n$ representing the reachability set of the plant (Σ) modelled by (1.4).

Assuming that the elements of $\overline{u}_N$ in (4.13) are either – 1 or 1, one easily ascertains that the range of R_N is a convex polyhedron whose vertices are defined by (4.11). Moreover, since R_N is surjective, it follows that for any state x_N belonging to (P) there exists at least one piecewise constant conrol input with the sampling period T satisfying constraint (4.10), which transfers the system state from the origin to x_N within the total time $t_f = NT$. ■

Remark 4.2

In practice, it is important to know the shape of (P) which depends on the numerical conditioning of R_N. The singular values of R_N give a natural characterization of the form of the reachability hyper-polyhedron (P). A considerable discrepancy between the largest and smallest singular values of R_N means a "flattened shape" of (P) corresponding to the used sampling and holding period T. In this context, the quality of the state controllability is reflected by the values of the angles between the hyper-polyhedron edges; further details will be given in Part III devoted to pole assignment problems. ■

The connection with the case of energy constrained control (discussed in 4.1) follows as a consequence of Theorem 4.1 and Theorem 4.5.

Corollary 4.3

A hyper-polyhedron (P) defined according to Theorem 4.5 is included in a hyper-ellipsoid whose semi-axes are

$$a_i = \sqrt{mN} v_i, \; i = \overline{1,n},$$

where $v_i, i=\overline{1,n},$ have the same meaning as in Theorem 4.1. ■

The effects of decreasing the sampling and holding period T are formulated as follows:

Theorem 4.6

Let (P) and (P') be two hyper-polyhedra defined according to Theorem

4.5 for the sampling and holding periods $T = t_f / N$, $N \geq \mu$ and $\nu \in \mathbf{N}$, $\nu \geq 2$, respectively.

Then:

$$(P) \subseteq (P')$$

and all the vertices of (P) are vertices of (P') as well.

Proof: Let $u_1, u_2, ..., u_N$ be any control sequence with the sampling and holding period T, satisfying constraint (4.10), or, equivalently, let us consider $\overline{u}_N \in \mathbf{R}^{Nm}$ satisfying (4.12), which transfers the system state from the origin to $x^* \in (P)$.

The same control sequence may be seen as a control sequence with the sampling and holding period T', i.e., $\overline{u}_{\nu N} \in \mathbf{R}^{\nu Nm}$, $u_{(k-1)\nu+1} = ... = u_{k\nu}$, $k = \overline{1, N}$, satisfying constraint (4.10) and transferring the system state from the origin to $x^* \in (P')$.

Using the notation E' and H' for the matrices of the discrete-time linear recursive and behavioural model (1.5) corresponding to T' and denoting by h'_i, $i = \overline{1, m}$, the i -th column of H', the vertices of (P') are given by:

$$\sum_{k=1}^{\nu N} \sum_{i=1}^{m} \sigma_{ki} (E')^{k-1} h'_i$$

with suitable $\sigma_{ki} \in \{-1, 1\}$ ensuring the convexity of (P').

According to the proof of the Theorem given in Appendix B, one may write:

$$\sum_{k=(l-1)\nu+1}^{l\nu} (E')^{k-1} h'_i = E^{l-1} h_i, \; 1 \leq l \leq N,$$

and consequently, those vertices of (P') obtained for σ_{ki} = constant, $k=\overline{(l-1)\nu+1, l\nu}$, are also vertices of (P). ■

Remark 4.3

By decreasing T , the form of the reachability hyper-polyhedron (P) becomes more advantageous, i.e., (P) is enlarged and includes new regions in $\mathbf{R}^n$. However, it is worth noticing that one cannot essentially improve the "quality" of (P) by decreasing T when the continuous-time model of the plant (Σ) is poorly state-controllable. ■

As a direct consequence of Theorem 4.6 and Corollary 4.1 one can state the following:

Corollary 4.4

A hyper-polyhedron (P) defined according to Theorem 4.5 is included in a hyper-ellipsoid whose semi-axes are given by the SVD of the Gram matrix:

$$\sqrt{mt_f}\int_0^{t_f} e^{At} BB^\tau e^{A^\tau t} dt. \qquad ■$$

The usefulness of linear non-recursive behavioural models consists in the possibility of undertaking direct studies (i.e., based on practical experiments) on the geometry of the reachability sets of the real plant (Σ); matrix R_N can be easily obtained from any linear non-recursive behavioural model with orthogonal test signals (X_N, U_N) associated with (Σ).

Consider a linear non-recursive behavioural model (X_N, U_N) corresponding to a sufficiently small value of the sampling and holding period T. The usefulness of linear non-recursive behavioural models consists in the possibility of undertaking direct studies (i.e., based on practical experiments) on the geometry of the reachability sets of the real plant (Σ); matrix R_N can be easily obtained from any linear non-recursive behavioural model with orthogonal test signals (X_N, U_N) associated with (Σ).

Let (Σ) be a plant described by the linear recursive behavioural model (1.4), which is completely state-controllable and the input signal of form (2.1) is constrained by (4.10), for the time horizon $t_f = NT$. Theorem 4.3 and Corollary 4.4 show that a fair approximation of the maximal reachability set is given by a hyper-ellipsoid whose semi-axes are:

$$a_i = (1/\alpha)\sqrt{mN}\,\overline{v}_i, \quad i = \overline{1,n}.$$

Vectors $\overline{v}_i$, $i = \overline{1,n}$, have the same meaning as in Theorem 4.3.

4.3. Effects of quantized control

In the vast majority of practical applications, the control input (2.1) is obtained as a quantized value vector supplied by D/A converters, i.e.,

$$u_k = qw_k, \; q \in \mathbf{Q}, \; w_k \in \mathbf{Z}^m. \tag{4.14}$$

The sampling and holding period $T > 0$ and the quantized increment q depend on the digital environment. Furthermore, the plant (Σ) is usually actuated by constrained inputs and that is why, without any loss of generality, it is assumed that (4.10), or equivalently (4.12), is satisfied.

If for system (1.4) the sampled and quantized control inputs satisfy (2.1), and (4.10) is used together with (4.14), then the reachability set represents a finite collection of points inside the reachability hyper-polyhedron (P) (corresponding to the same sampling and holding period $T > 0$, see Theorem 4.1), organized as a grid (G). For a period $T = (t_f / N)$, $N \in \mathbf{N}$, $N \geq \mu$, this grid is defined as a consequence of (2.1), (2.3) and (4.14), by the points:

$$qR_N\overline{w}_N \in (P),\ q \in \mathbf{Q},\ \overline{w}_N \in \mathbf{Z}^{Nm}. \tag{4.15}$$

The system state can be transferred from the origin to any desired final state $x^* \in (P)$ with the accuracy:

$$\left\| x(t_f) - x^* \right\|_2 = \min_{\overline{w}_N \in \mathbf{Z}^{Nm}} \left\| qR_N \overline{w}_N - x^* \right\|_2 \tag{4.16}$$

It is a commonplace to note that by decreasing the quantizer increment q one obtains a better accuracy in (4.16). Indeed, for two quantizer increments q and q', satisfying $q = \nu q'$, $\nu \in \mathbf{N}$, $\nu \geq 2$, the accuracy in reaching the desired final state $x^* \in (P)$ is characterized by:

$$\min_{\overline{w}_N \in \mathbf{Z}^{Nm}} \left\| qR_N\overline{w}_N - x^* \right\|_2 = \min_{\overline{w}_N \in \mathbf{Z}^{Nm}} \left\| q'R_N(\nu\overline{w}_N) - x^* \right\|_2 \geq$$
$$\geq \min_{\overline{w}'_N \in \mathbf{Z}^{Nm}} \left\| q'R_N\overline{w}'_N - x^* \right\|_2 .$$

From the practical point of view, for usual applications and environments, decreasing q is less convenient than decreasing T and that is why the following result from Pastravanu and Voicu (1992c) becomes very useful.

Theorem 4.7

Let (G) and (G') denote two grids corresponding to (4.15) for the sampling and holding periods $T = (t_f / N)$ and $T' = (t_f / \nu N)$, $\nu \in \mathbf{N}$, $\nu \geq 2$, respectively, which belong to the hyper-polyhedron (P), defined in accordance with Theorem 4.5.

Then there exists the following inclusion:

$$(G) \subseteq (G')$$

which allows one to increase the accuracy in (4.16).

Proof: For (G') using (4.15) *mutatis mutandis*, one can write:

$$qR_{\nu N}\overline{w}_{\nu N} \in (P),\ \overline{w}_{\nu N} \in \mathbf{Z}^{\nu Nm}, \tag{4.17}$$

which shows that all points of (G) are obtainable from (4.17) with

$$w_{(j-1)\nu+1} = \ldots = w_{j\nu},\ j = \overline{1,N}.$$

At the same time, it is a simple matter to see that:

$$\min_{\overline{w}_{\nu N}\in\mathbf{Z}^{\nu Nm}} \left\|qR_{\nu N}\overline{w}_{\nu N} - x^*\right\|_2 \leq$$
$$\leq \min_{\overline{w}_{N}\in\mathbf{Z}^{Nm}} \left\|qR_{\nu N}(M_{\nu N}\overline{w}_{N}) - x^*\right\|_2 = \min_{\overline{w}_{N}\in\mathbf{Z}^{Nm}} \left\|qR_{N}\overline{w}_{N} - x^*\right\|_2,$$

which completes the proof. Matrix $M_{\nu N}$ has the same significance as the proof of Theorem 4.2; see also Appendix B. ■

Remark 4.4

When using the sampled and quantized control input (2.1), if the quantizer increment cannot be decreased, then Theorem 4.7 offers a possibility to improve the accuracy in reaching the desired final state, by decreasing the value of the sample period. ■

PART III.

FEEDBACK CONTROL DESIGN USING LINEAR NON-RECURSIVE BEHAVIOURAL MODELS

CHAPTER 5

POLE ASSIGNMENT

5.1. Single-input case

According to Ackermann (1977), most of the real systems (Σ), described by the linear recursive behavioural model (1.4), have in their discrete-time pendants (1.5) cyclic matrices E, and for almost every $d \in \mathbf{R}^m$ the pair (E, h), with $h = Hd$, is completely state-controllable. In other words, using a suitable combination of the input channels, i.e.,

$$u(t) = dv(t), \; v \in \mathbf{R},$$

models (1.4) and (1.5) as single-input control systems:

$$\dot{x}(t) = Ax(t) + bv(t), \quad \text{with } b = Bd,$$

and

$$x(t+T) = Ex(t) + hv(t), \quad \text{with } h = H\,d,$$

respectively, become completely state-controllable systems (for the latter, the condition in Theorem 3.1 should be fulfilled).

This is the starting hypothesis in determining the state feedback controller for pole assignment in Pastravanu and Voicu (1991) and for our further developments.

The procedure of state controllability testing based on linear non-recursive behavioural models (see Chapter 3) allows the determination of an appropriate $d \in \mathbf{R}^m$ so that the state controllability degrees (see Chapter 4) of the single-input control system reflect good state controllability properties.

In order to avoid the introduction of new variables and notations for the control input, let us use u instead of v, so that the single-input models considered in the remainder of this paragraph will be written as:

$$\dot{x}(t) = Ax(t) + bu(t),\ t \in \mathbf{R}_+,\ x \in \mathbf{R}^n,\ u \in \mathbf{R}, \qquad (5.1)$$

and

$$x(t+T) = Ex(t) + hu(t),\ t \in [kT,(k+1)T), \quad k \in \mathbf{N}_0, \ (5.2)$$

respectively, where:

$$E = e^{At},\ h = \int_0^T e^{At} dt\, b.$$

Behavioural models (5.1) and (5.2) must be understood as special forms of (1.4) and (1.5) respectively, not only for the case $m = 1$, but also for $m > 1$ and for an appropriate linear combination of input channels as presented above.

The desired closed-loop dynamics is achieved using the state feedback controller:

$$u(t) = f^{\tau} x(kT),\ t \in [kT,(k+1)T),\ k \in \mathbf{N}_0, \qquad (5.3)$$

with $f \in \mathbf{R}^n$, which assigns for the closed-loop system (5.2), (5.3) the spectrum:

$$\sigma(E + hf^{\tau}) = Z \triangleq \{z_i \in \mathbf{C},\ i = \overline{1,n};\ |z_i| < 1\}. \qquad (5.4)$$

Equivalently, the closed-loop characteristic polynomial is:

$$P(z) = \prod_{i=1}^{n} (z - z_i) \triangleq \sum \alpha_j z^j,\ \alpha_n = 1, \qquad (5.5)$$

according to the design problem (**P**) stated in 1.1.

In order to present a solution based exclusively on linear non-recursive behavioural models, first let us reformulate Theorem 1.1 for the single-input case.

Theorem 5.1

For any $P(z)$ defined by (5.5), and associated with Z defined by (5.4), a state feedback controller (5.3) which ensures for the closed-loop system (5.1), (5.3) the pole placement into set Z is given by the unique solution of the algebraic linear system:

$$[x^q(t_0^q)]^\tau f = -[1/\Pr_p(E^{n-1}h)]\sum_{j=0}^{n}\alpha_j \Pr_p x_{fr}^q(t_0^q + jT). \quad (5.6)$$

The initial conditions $x^q(t_0^q)$, $q=\overline{1,n}$, are arbitrary and linearly independent, p denotes the unity vector of the orthogonal direction on:

$$\Pi = \mathrm{Span}_{\mathbf{R}}(h, Eh, ..., E^{n-2}h)$$

and $\Pr_p w$ is the orthogonal projection of $w \in \mathbf{R}^n$ along p. ■

Now one can prove the following result:

Theorem 5.2

Given an arbitrary extended linear non-recursive behavioural model $(X_{2n},...,X_n,U_n)$ of the real plant (Σ), a state feedback controller (5.3) which ensures the closed-loop pole placement into any set Z (associated with polynomial (5.5)), is the unique solution of the algebraic linear system:

$$\begin{bmatrix} X_n^\tau & 0 \\ \sum_{j=0}^{n}\alpha_j X_{n+j}^\tau & X_n^\tau \end{bmatrix}\begin{bmatrix} g \\ f \end{bmatrix} = \begin{bmatrix} u_1^1 ... u_1^n & 0...0 \end{bmatrix}^\tau, \quad (5.7)$$

where u_1^i, $i=\overline{1,n}$, are the elements of the last row of matrix $U_n \in \mathbf{R}^{n\times n}$, and g denotes an auxiliary vector defined as:

$$g = \varepsilon p,\ \varepsilon \in \mathbf{R},\ \varepsilon \,\mathrm{Pr}_p(E^{n-1}h) = 1. \tag{5.8}$$

Proof: From the first n equations of (5.7), with (2.5) for $j = n$ and $m = 1$, one can write:

$$U_n^\tau R_n^\tau g = [u_1^1 \ldots u_1^n]^\tau,$$

which uniquely yields:

$$R_n^\tau g = [0 \ldots 1]^\tau.$$

In fact, the latter is equivalent to (5.8) and offers the possibility to determine the auxiliary vector g.

The last n equations of (5.7) can be globally written as:

$$X_n^\tau f = -\sum_{j=0}^{n} \alpha_j X_{n+j}^\tau g. \tag{5.9}$$

Each equation in (5.9) has the form:

$$(x_n^q)^\tau f = -\sum \alpha_j (x_{n+j}^q)^\tau [1 / \mathrm{Pr}_p(E^{n-1}h)] p,\ q = \overline{1, n},$$

or equivalently:

$$(x_n^q)^\tau f = -[1/\mathrm{Pr}_p(E^{n-1}h)] \sum_{j=0}^{n} \alpha_j \,\mathrm{Pr}_p \, x_{n+j}^r,\ q = \overline{1, n}. \tag{5.10}$$

Now it is obvious that taking $x_n^q = x^q(t_0^q)$, $q = \overline{1, n}$, the algebraic linear system (5.10) is equivalent to (5.6), i.e., Theorem 5.2 is equivalent to Theorem 5.1, because X_n is non-singular as a consequence of the complete state-controllability, which was assumed from the very beginning.

The effective solution for f follows by separately solving the two algebraic subsystems contained in (5.7), finally resulting in:

$$f^{\tau} = -[u_1^1 \dots u_1^n](X_n)^{-1}\sum_{j=0}^{n}\alpha_j X_{n+j}(X_n)^{-1}. \qquad \blacksquare$$

Remark 5.1

From (5.7) it is obvious that one can obtain small entries in f (i.e., small feedback gain factors) using a sampling and holding period $T > 0$ which allows one to increase the minimal singular value of matrix X_n and to decrease the maximal singular value of matrix $\sum_{j=0}^{n}\alpha_j X_{n+j}$

Considering now the equivalence stated by Theorem 2.1 (see Chapter 2), the following result can be easily formulated.

Corollary 5.1

In terms of Theorem 5.2, but for matrices D_k, $k = \overline{0,n}$, obtained from the step response of the real plant (Σ), f is the unique solution of:

$$\begin{bmatrix} D_0^{\tau} & 0 \\ \sum_{j=0}^{n}\alpha_j D_j^{\tau} & D_0^{\tau} \end{bmatrix}\begin{bmatrix} g \\ f \end{bmatrix} = \begin{bmatrix} 0 \dots \alpha & 0 \dots 0 \end{bmatrix}^{\tau}. \blacksquare \qquad (5.11)$$

Remark 5.2

According to the geometric significance of g, the first n equations pertaining to system (5.11), i.e.,

$$D_0^{\tau} g = [0 \dots \alpha]^{\tau},$$

can be replaced by

$$\Delta_0^{\tau} g = [0 \dots \alpha]^{\tau},$$

where the first $n-1$ columns of matrix Δ_0 are orthogonal (i.e., an orthogonal basis of Π, see Theorem 5.1) and the last column of Δ_0 is taken directly from D_0. This new form is more advantageous from the numerical point of view. ■

5.2. Multi-input case

For the next development using the extended linear non-recursive behavioural model $(X_{2n},...,X_n,U_n)$, the following matrices are defined:

$\mathcal{X}_n \in \mathbf{R}^{n\times n}$ which contains the first n linearly independent columns of matrix X_n;

$\mathcal{X}_{n+j} \in \mathbf{R}^{n\times n}$, $j=\overline{1,n},$ which contains those columns of X_{n+j}, $j=\overline{1,n},$ the positions of which correspond to the positions already selected in X_n for constructing $\mathcal{X}_n$.

By $\mathcal{R}_n \in \mathbf{R}^{n\times n}$ one denotes the matrix obtained from R_n (corresponding to (2.6) for $j=n$) by extracting the first n linearly independent columns and by ordering, in the last m positions, the columns which represent the vectors $E^{n_1-1}h_1,...,$ $E^{n_m-1}h_m$; and vectors h_i, $i=\overline{1,m},$ denote the columns of H, i.e., the first m columns of R_n.

As in the single-input case, the real plant (Σ) is assumed as completely state-controllable, and may be described by linear recursive behavioural models (1.4) and (1.5) (for the latter, the condition in Theorem 3.1 must be fulfilled). Consider a linear extended non-recursive behavioural model resulting from the experimentally acquired data as presented in Chapter 2. Moreover, assume that by appropriate state controllability tests, the Kronecker invariants n_i, $i=\overline{1,m},$ and the "α-parameters" and "β-parameters" (using the original notation in Popov, 1972) of matrix R_n have been determined (see Chapter 3).

Theorem 5.3

Given an arbitrary extended linear non-recursive behavioural model $(X_{2n},\ldots,X_n,U_n)$ with orthogonal test signals, associated with plant (Σ), a state feedback controller (1.6) which ensures the closed-loop pole placement into any set Z defined by (1.7), (or equivalently by polynomial $P(z)$ defined by (1.23) with (1.12)), is given by:

$$F = MK. \tag{5.12}$$

Matrix M has been defined by (1.14) and

$$K^{\tau} = [k_1,\ldots,k_m] \in \mathbf{R}^{n\times m} \tag{5.13}$$

is built with the unique solutions $k_i \in \mathbf{R}^n$, $i=\overline{1,m}$, of the following m matrix equations:

$$\begin{bmatrix} \mathcal{R}_n^{\tau} & 0 \\ \sum_{j=0}^{n_i} \alpha_{ij} X_{n+j}^{\tau} & X_{n+j}^{\tau} \end{bmatrix} \begin{bmatrix} g_i \\ k_i \end{bmatrix} = [0\ldots0 \underbrace{1}_{n-m+i} 0\ldots0]^{\tau},\ i=\overline{1,m}, \tag{5.14}$$

where the auxiliary vectors $g_i, \in \mathbf{R}^n$, $i=\overline{1,m}$, have been defined by (1.11).

Proof: For each i, $i=\overline{1,m}$, the first n equations in (5.14) are just the definition equations (1.11) for auxiliary vectors g_i, $i=\overline{1,m}$.

For each i, $i=\overline{1,m}$, the last n scalar equations of (5.14) can be rewritten in the form given by (1.30) (see Theorem 1.1), i.e., Theorems 5.3 and 1.1 are equivalent. ■

Denoting by n_δ the controllability index of the linear recursive behavioural model (1.5), i.e., $n_\delta = \max_{i=\overline{1,m}} n_i$, the state feedback control (1.6) can be determined so that the closed-loop minimal polynomial has the degree n_δ.

The following result is also a direct consequence of Corollary 1.1.

Corollary 5.2

A state feedback control (1.6) which ensures for the closed-loop system the minimal polynomial $P_{\min}(z)$, with the degree n_δ, is given by (5.12)–(5.14) where $P_{\min}(z)$, factorized according to (1.23) and (1.12), satisfies $P_\delta(z) = P_{\min}(z)$, $P_i(z) \mid P_\delta(z)$ for $i = \overline{1,m}$. ■

Remark 5.3

According to Remark 1.2, the necessity of choosing the sampling and holding periods $T > 0$ which ensure small values for $\Pr_{p_i}(E^{n_i - 1} h_i)$, $i = \overline{1,m}$, becomes obvious. As $1 / r_n = \left\| \mathcal{R}_n^{-1} \right\|_2$ represents a majorant for $\Pr_{p_i}(E^{n_i - 1} h_i)$, $i = \overline{1,m}$, ($r_n > 0$ denoting the minimal singular value of matrix R_n), the result is that, in principle, by choosing $T > 0$ so that r_n is maximized, one can ensure small values for the entries of matrix K. At the same time, it is worth noticing that small values of the β -parameters lead to small values for $\left\| M \right\|_2$ so that, generally speaking, small gain factors in matrix F are guaranteed. According to section 3.4, all these considerations may be easily reformulated in terms of the state controllability quality or state controllability degrees. ■

Remark 5.4

In Pastravanu and Voicu (1992b), one can find a purely algebraic proof of Theorem 5.3 derived from Theorem 4 in Ackermann (1984), without any reference to the geometric interpretation presented in 1.3. ■

Relying on the equivalence established in Theorem 2.1, the algebraic linear system (5.14) may also be expressed in terms of the step response of the real plant (Σ). Denoting by $\mathcal{D}_0 \in \mathbf{R}^{n\times n}$ the matrix which contains the first n linearly independent columns of D_0 (see Theorem 2.1) and by $\mathcal{D}_j \in \mathbf{R}^{n\times n}$, $j=\overline{1,n}$, the matrices containing those columns of D_j, $j=\overline{1,n}$, which correspond to the column positions already selected from D_0 for constructing $\mathcal{D}_0$, it is commonplace to note that:

$$\mathcal{X}_{n+j} = \mathcal{D}_j, \; j=\overline{0,n}.$$

Corollary 5.3

In terms of Theorem 5.3, but for matrices $\mathcal{D}_j$, $j=\overline{0,n}$, obtained from the step response of the real plant (Σ) (see Chapter 2), matrix F is given by (5.12) and (5.13), where k_i, $i=\overline{1,m}$, are the unique solutions of the following matrix equations:

$$\begin{bmatrix} \mathcal{R}_n^{\tau} & 0 \\ \sum_{j=0}^{n_i} \alpha_{ij}\mathcal{D}_j^{\tau} & \mathcal{D}_0^{\tau} \end{bmatrix}\begin{bmatrix} g_i \\ k_i \end{bmatrix} = [0 \ldots 0 \underbrace{1}_{n-m+i} 0\ldots 0]^{\tau},\; i=\overline{1,m}. \blacksquare \quad (5.15)$$

Note that the same equivalence (stated in Theorem 2.1) allows a simplified computation of matrix $\mathcal{R}_n$ since $R_n = \alpha^{-1} D_0$.

5.3. Numerical aspects

Since the experimentally acquired data are altered by measurement errors, all the computations required by the pole placement method should be done using perturbed matrices in the extended linear non-recursive behavioural model, which represents the sole information about the behaviour of the real plant (Σ). As a matter of fact, for many environments, the quantization errors are much more important than the truncation or round-off errors, because the number of significant decimal digits guaranteed by A/D and D/A converters is usually smaller than the one involved in the arithmetic of the simple precision floating point.

The perturbation theory in linear algebra (see e.g. Golub and Van Loan (1983)) can offer upper bounds for the relative errors that occurred in determining the state feedback controller (1.6) by means of the proposed pole assignment method presented in Pastravanu and Voicu (1992a) and Pastravanu (1992).

Denote by M_i, $i=\overline{1,m}$, the coefficient matrices of the algebraic linear system (5.14), by $s_l(\bullet)$, $l=\overline{1,r}$, with $s_1(\bullet)\geq\ldots\geq s_r(\bullet)>0$ the singular values of the $r\times r$ matrix $(\bullet)$, and by ε the error due to the measurement.

Theorem 5.4

For any extended linear non-recursive behavioural model, with orthogonal test signals, of the real plant (Σ), the relative errors e_i, $i=\overline{1,m}$, (expressed in terms of the spectral norm) committed when solving system (5.14) with respect to the actual values of the unknown parameters of (Σ), satisfy:

$$e_i \leq n\varepsilon\sqrt{1+\alpha^{-2}+\left(\sum_{j=0}^{n_i}\left|\alpha_{ij}\right|\right)^2}\Big/ s_{2n}(M_i),\ i=\overline{1,m}. \quad (5.16)$$

Proof: Denoting by ΔM_i the additive perturbation of M_i, according to (3.4), one can write:

$$\left\|\Delta M_i\right\| \leq \sqrt{n^2(\varepsilon/\alpha)^2+n^2\varepsilon^2+\left(\sum_{j=0}^{n_i}\left|\alpha_{ij}\right|\right)^2 n^2\varepsilon^2}$$

and consequently, the sensitivity of the square linear system (5.14) gives the majorants (5.16). ■

An adequate approach must pertain to the practical aspects involved by condition (5.16) in the context of experimentally acquired matrices X_{n+j},

$j = \overline{0,n}$, in the extended linear non-recursive behavioural model, which ensures, as in Pastravanu and Voicu (1992a),

$$s_n(X_n) \gg n\varepsilon, \tag{5.17}$$

assuming that the considered system is completely state-controllable.

In the sequel, it will be shown that for those real plants (Σ) and/or those sampling and holding periods $T > 0$ for which one can experimentally obtain well-conditioned matrices X_n, i.e., (5.17) is fulfilled, the majorants (5.16) have quite small values.

Indeed, by using the factorization:

$$M_i = \begin{bmatrix} \mathcal{R}_n^\tau & 0 \\ 0 & \mathcal{X}_n^\tau \end{bmatrix} N_i \quad \text{with} \quad N_i = \begin{bmatrix} I_n & 0 \\ [P_i(E)]^\tau & I_n \end{bmatrix},$$

which may be easily proved from (1.12) and (2.12), according to the properties of singular values, one concludes that:

$$s_{2n}(M_i) \geq \min[s_n(\mathcal{X}_n), s_n(\mathcal{R}_n)]\ s_{2n}(N_i). \tag{5.18}$$

Now for singular values $s_j(N_i)$, $j = \overline{1,2n}$, one can prove the following result.

Theorem 5.5

Singular values $s_j(N_i)$ and $s_{2n-j+1}(N_i)$, $j = \overline{1,n}$, are the roots of the equation:

$$s^2 - \{2 + [s_j(P_i(E))]^2\}s + 1 = 0,\ j = \overline{1,n}. \tag{5.19}$$

Proof: The singular values of M_i are the roots of the equation:

$$\det(I_{2n}s - N_i^\tau N_i) = 0.$$

By a suitable matrix partitioning one can write:

$$\det(I_{2n} - N_i^\tau N_i) = \det\begin{bmatrix} I_n s - I_n & -P_i^\tau(E) \\ -P_i(E) & I_n s - I_n - P_i(E)P_i^\tau(E) \end{bmatrix}$$

which, according to Schur's formula, can be developed as follows:

$$\begin{aligned} \det(I_{2n}s - N_i^\tau N_i) &= \det(I_n s - I_n)\det[I_n s - I_n - P_i(E)P_i^\tau(E) - \\ &- P_i(E)(I_n s - I_n)^{-1} P_i^\tau(E)] = \\ &= (s-1)^n \det[I_n s - I_n - P_i(E)P_i^\tau(E) - \\ &- (s-1)^{-1} P_i(E)P_i^\tau(E)] = \\ &= \det[(s-1)(I_n s - I_n - P_i(E)P_i^\tau(E) - \\ &- P_i(E)P_i^\tau(E)] = \\ &= \det[(s-1)^2 I_n - P_i(E)P_i^\tau(E)s] = \\ &= s^n \det[(s-1)^2/s - P_i(E)P_i^\tau(E)] = \\ &= s^n \prod_{j=1}^n \{(s-1)^2/s - [s_j(P_i(E))]^2\} = \\ &= \prod_{j=1}^n \{s^2 - \{2 + [s_j(P_i(E))]^2\}s + 1\}. \end{aligned}$$

This result proves the theorem. ■

According to Theorem 5.5, it results that,

$$s_1(N_i) = s_{2n}^{-1}(N_i)$$

and moreover, $s_1(N_i)$ and $s_{2n}(N_i)$ are the roots of the equation:

$$s^2 - (2 + \|P_i(E)\|^2)s + 1 = 0.$$

It follows that $1/(2+\|P_i(E)\|^2)$ represents a minorant for $s_{2n}(N_i)$. Therefore it results that:

$$\frac{1}{s_{2n}(M_i)} \leq \frac{2+\|P_i(E)\|^2}{\min[s_n(\mathcal{X}_n), s_n(\mathcal{R}_n)]}.$$

Roughly speaking, this means that for $\|P_i(E)\|$ that is sufficiently small and for well-conditioned matrices $\mathcal{X}_n$ and $\mathcal{R}_n$ (in fact (5.17) fulfilled), Theorem 5.4 gives useful upper bounds for the relative errors e_i, $i=\overline{1,m}$.

CHAPTER 6

STRUCTURALLY STABLE COMPENSATOR SYNTHESIS

6.1. Single-input single-output case

Consider the real plant (Σ), which may be described by the linear recursive behavioural model (5.1) and by its discrete-time pendant (5.2), with the single-output:

$$y(t) = c^{\tau} x(t), \; y \in \mathbf{R}. \tag{6.1}$$

For this system, the problem $(\mathbf{P}')$ (see Chapter 1) is considered, i.e., one must design a compensator which, by the appropriate pole assignment, ensures the desired closed-loop dynamics and zero steady-state error for the exogenous step.

The compensator incorporates an internal discrete-time model of the exogenous variable as described in Wonham (1979):

$$x_{n+1}(t+T) = x_{n+1}(t) - y(t) + r(t), \; t = kT, \; k \in \mathbf{N}_0, \tag{6.2}$$

where $r(t$ denotes the set point.

Under these circumstances the state vector is augmented as follows:

$$\tilde{x}(kT) = \begin{bmatrix} x(kT) \\ x_{n+1}(kT) \end{bmatrix} \in \mathbf{R}^{n+1} \tag{6.3}$$

and the state feedback becomes:

$$u(t) = \tilde{f}^{\tau} \tilde{x}(kT), \; t \in [kT, (k+1)T), \tag{6.4}$$

where:

$$\tilde{f} = \begin{bmatrix} f \\ f_{n+1} \end{bmatrix} \in \mathbf{R}^{n+1} \tag{6.5}$$

also includes the gain factor corresponding to the state variable x_{n+1}.

The desired closed-loop dynamics of the system (5.1), (6.1), (6.2), (6.4) may be achieved by the pole placement into the set:

$$\tilde{Z} = \{z_i \in \mathbf{C},\ i = \overline{1, n+1},\ |z_i| < 1\}. \tag{6.6}$$

First let us prove the following preparatory result.

Proposition 6.1

For those sampling and holding periods $T > 0$ for which the discrete-time system (5.2) conserves the complete state-controllability of the continuous-time system (5.1), the augmented system (5.2), (6.1), (6.2) is also completely state-controllable if and only if:

$$c^{\tau} A^{-1} b \neq 0.$$

Proof: In Porter and Power (1970) it is proved that the continuous-time system (5.1) with (6.1), augmented with the internal continuous-time behavioural model of the exogenous variable:

$$\dot{x}_{n+1} = -y + r \tag{6.7}$$

is completely state-controllable if and only if $G(0)=c^{\tau}(-A)^{-1}b\neq 0$ where $G(s)=c^{\tau}(I_n s-A)^{-1}b$ is the transfer function (from u to y) of system (5.1), (6.1). For $s = 0$, one obtains the condition of the proposition. Moreover, equation (6.7) introduces a zero eigenvalue, the others being the eigenvalues of matrix A. Thus, the values of T which conserve the complete state-controllability of system (5.2), ensure the same property for the system (5.2), (6.1), (6.2) according to Theorem 3.1. ■

Remark 6.1

Condition $c^{\tau}A^{-1}b \neq 0$ is equivalent to $y_{st} = \lim_{t\to\infty} y(t) \neq 0$ for $u(t) =$ $= \alpha\sigma(t)$,

$$y_{st} = \lim_{s\to 0} sG(s)(\alpha / s) = \alpha G(0) = \alpha c^{\tau}(-A)^{-1}b \neq 0.$$

Clearly, it is easy to verify experimentally that $y_{st} \neq 0$. ■

To get the necessary data, one experimentally acquires the first $2n+2$ values of the state vector x obtained for a step input $u(t) = \alpha\sigma(t)$

$$\tilde{x}_{\sigma}(T), \tilde{x}_{\sigma}(2T), ..., \tilde{x}_{\sigma}(2n+2), \tag{6.8}$$

With (6.8) one builds the matrix:

$$\tilde{D} = [\tilde{d}_1, ..., \tilde{d}_{2n+2}] \in \mathbf{R}^{(n+1)\times(2n+2)} \tag{6.9}$$

using the procedure given by (2.15), (2.16) (*mutatis mutandis*). From $\tilde{D}$ one extracts matrices $\tilde{D}_j \in \mathbf{R}^{(n+1)\times(n+1)}, j = \overline{0, n+1}$, containing $n+1$ consecutive columns beginning with the j-th column in $\tilde{D}$.

Now defining the polynomial:

$$\tilde{P}(z) = \prod_{i=1}^{n+1}(z - z_i) = \sum_{j=0}^{n+1} \beta_j z^j,\ \beta_{n+1} = 1, \tag{6.10}$$

associated with the set $\tilde{Z}$, according to Corollary 5.1 and Proposition 6.1 one can formulate the following result.

Theorem 6.1

In terms of Proposition 6.1 and Corollary 5.1 (*mutatis mutandis*), $\tilde{f}$ is the unique solution of:

$$\begin{bmatrix} \tilde{D}_0^\tau & 0 \\ \sum_{j=0}^{n+1} \beta_j \tilde{D}_j^\tau & \tilde{D}_0^\tau \end{bmatrix} \begin{bmatrix} \tilde{g} \\ \tilde{f} \end{bmatrix} = [0...0 \underbrace{\alpha}_{n+1} 0...0]^\tau, \qquad (6.11)$$

where $\tilde{g} \in \mathbf{R}^{n+1}$ is an auxiliary vector. ■

6.2. Self-learning procedure

In some situations, it is possible to build a linear non-recursive behavioural model from the normal functioning of the real plant (Σ), described by (5.1), i.e., for an arbitrary piecewise constant input signal, with the sampling and holding period $T > 0$ (according to Theorem 3.1).

Consider the real plant (Σ) with a zero initial condition and the arbitrary single-input sequence:

$$u_0, u_1, ..., u_{2n-1}, \; u_0 \neq 0.$$

One acquires the following sequence of the state vector:

$$x_a(T), x_a(2T), ..., x_a(2nT),$$

and computes, recursively, the vectors:

$$d_1 = u_0^{-1} x_a(T)$$

$$........ \qquad (6.12)$$

$$d_i = u_0^{-1}[x_a(iT) - \sum_{k=1}^{i-1} u_k d_{i-k}], \; i = \overline{2, 2n},$$

which constitute the columns of the matrix:

$$D = [d_1, d_2, ..., d_{2n}] \in \mathbf{R}^{n \times 2n}.$$

Now one extracts from D the matrices $D_j \in \mathbf{R}^{n \times n}, j = \overline{0, n},$ containing n

consecutive columns of D beginning with the $j+1$-st column in D.

Proposition 6.2

Matrices $X_{n+j}, j=\overline{0,n}$, of the extended linear non-recursive behavioural model $(X_{2n},\ldots,X_n,U_n)$ with $U_n=\alpha I_n, \alpha>0$, satisfy the following equalities:

$$X_{n+j}=\alpha D_j,\ j=\overline{0,n}. \tag{6.13}$$

Proof: From $U_n=\alpha I_n$, (2.6) and (2.12) it follows that:

$$X_{n+j}=\alpha E^j R_n,\ j=\overline{0,n}. \tag{6.14}$$

On the other hand, one obtains:

$$d_1=h,\quad d_2=Eh.$$

Suppose that:

$$d_i=E^{i-1}h.$$

Taking into account (2.2) it follows that:

$$x_a((i+1)T)=[h\ldots E^i h][u_i\ldots u_0]^\tau,$$

and from (6.12) it results that:

$$d_{i+1}=(1/u_0)\left[\sum\nolimits_{k=0}^{i} E^{i-k}hu_k-\sum\nolimits_{k=1}^{i} E^{i-k}hu_k\right],$$

i.e.,

$$d_{i+1}=(1/u_0)E^i hu_0=E^i h.$$

Thus, it has been proved by complete induction that:

$$d_i = E^{i-1}h,\ i=\overline{1,n},$$

and consequently, one may write:

$$D_j = [E^j h, E^{j+1}h, ..., E^{j+n-1}h] = E^j R_n,\ j=\overline{0,n}. \quad (6.15)$$

Now (6.13) results from (6.14) and (6.15). ■

This result makes it possible to use the normal functioning of the real plant (Σ) in order to construct the matrices $X_{n+j},\ j=\overline{0,n},$ of the extended linear non-recursive behavioural model $(X_{2n},...,X_n,U_n)$, with $U_n = \alpha I_n$. Theorem 5.2 opens a new perspective for compensator synthesis which may be approached as a self-learning procedure.

Let us consider the problem solved in section 6.l.

Assume that the closed-loop plant (Σ) with (6.1), (6.2), (6.4) is functioning around the operating point $\tilde{x}_0$ which is considered as the origin. For any variation of the set point $r(t)$ one acquires the first $2n+2$ values of the extended state vector, i.e.,

$$\tilde{x}(T), \tilde{x}(2T), ..., \tilde{x}((2n+2)T)$$

and the single-input control sequence:

$$u(0),...,u((2n+1)T),\ u(0)\neq 0.$$

With these acquired data one computes recursively:

$$\tilde{d}_1 = u^{-1}(0)\tilde{x}(T)$$

....

$$\tilde{d}_i = u^{-1}(0)[\tilde{x}(iT) - \sum_{k=1}^{i-1} u(kT)\tilde{d}_{i-k}],\ i=\overline{2,2n+2},$$

which constitute the columns of the matrix:

$$\tilde{D} = [\tilde{d}_1, \ldots, \tilde{d}_{2n+2}] \in \mathbf{R}^{(n+1)\times(2n+2)},$$

One extracts from $\tilde{D}$ the matrices $\tilde{D}_j \in R^{(n+1)\times(n+1)}$, $j = \overline{1, n+1}$, containing $n+1$ consecutive columns beginning with the $j+1$-st column in $\tilde{D}$.

Now, according to Theorem 6.1 and Proposition 6.2 (*mutatis mutandis*), one can obtain the state feedback gain $\tilde{f}$ directly from the closed-loop operation of (Σ), as constructed by a self-learning procedure.

Theorem 6.2

In terms of Theorem 6.1, the state feedback control (6.4) which ensures the closed-loop pole placement into the set $\tilde{Z}$, defined by (6.6) (or, equivalently, according to the polynomial $\tilde{P}(z)$ in (6.10)), is given by the unique solution of the following linear algebraic system:

$$\begin{bmatrix} \tilde{D}_0^\tau & 0 \\ \sum_{j=0}^{n+1} \beta_j \tilde{D}_j^\tau & \tilde{D}_0^\tau \end{bmatrix} \begin{bmatrix} \tilde{g} \\ \tilde{f} \end{bmatrix} = [0 \ldots 0 \underset{n+1}{\underset{\downarrow}{1}} 0 \ldots 0]^\tau$$

where $\tilde{g} \in \mathbf{R}^{n+1}$ is an auxiliary vector. ■

6.3. Multi-input multi-output case

Consider the real plant (Σ), described by the linear recursive behavioural model (1.4) and by its discrete-time pendant (1.5) with $m = p$.

We are going to refer to problem **(P')** presented in Chapter 1.

The structurally stable compensator incorporates an internal discrete-time model of the exogenous variable (Wonhanm, 1979):

$$x_e(t+T) = I_m x_e(t) - y(t) + r(t),\ t = kT,\ k \in \mathbf{N}_0, \quad (6.16)$$

where $x_e \in \mathbf{R}^m$ denotes the state vector of the internal behavioural model and $r \in \mathbf{R}^m$ is the set point vector.

Under these circumstances, the state vector of the whole augmented system is:

$$\tilde{x}(kT) = \begin{bmatrix} x(kT) \\ x_e(kT) \end{bmatrix} \in \mathbf{R}^{n+m} \quad (6.17)$$

and the state feedback becomes:

$$u(t) = \tilde{F}\tilde{x}(kT),\ t \in [kT, (k+1)T), \quad (6.18)$$

where:

$$\tilde{F} = \begin{bmatrix} F \\ F_e \end{bmatrix} \in \mathbf{R}^{m \times (n+m)} \quad (6.19)$$

also includes the gain matrix F_e corresponding to the state vector x_e.

The desired closed-loop dynamics of system (1.4), (6.16), (6.18) may be achieved by the pole placement into the set:

$$\tilde{Z} = \{z_i \in \mathbf{C},\ i = \overline{1, n+m};\ |z_i| < 1\}. \quad (6.20)$$

First let us prove a result referring to the state controllability of the system (1.4), (6.16), by referring to the discrete-time dynamics.

Let $y_{st_i} \in \mathbf{R}^m$, $i = \overline{1,m}$, be the steady state output vector of the open-loop system (1.4) for a step input applied to the i -th channel:

$$u_i(t) = \alpha \sigma_i(t),\ i = \overline{1,m}, \quad (6.21)$$

where:

$$\sigma_i(t) = [0...0\ \underbrace{1}_{i}\ 0...0]^{\tau} \in \mathbf{R}^m,\ t \in \mathbf{R}_+, \tag{6.22}$$

and $\alpha > 0$ represents an appropriate amplitude ensuring the accuracy of the measurement of y_{st_i}.

Let

$$Y_{st} = [y_{st_1} ... y_{st_m}] \in \mathbf{R}^{m \times m} \tag{6.23}$$

be the matrix built with y_{st_i}, $i = \overline{1, m}$.

Proposition 6.3

For those sampling and holding periods $T > 0$ for which the discrete-time behavioural system (1.5) preserves the complete state-controllability of the continuous-time behavioural system (1.4), the augmented system (1.5), (6.16) is also completely state-controllable if and only if:

$$\det Y_{st} \neq 0.$$

Proof: This is based on the general condition provided by (Porter and Power, 1970) meaning $\text{rank} G(0) = m$, where $G(s) = C(I_n s - A)^{-1} B$ denotes the transfer function matrix. Note that this represents the generalization of the condition in Proposition 6.1 which refers to the single-input single-output case. ■

For building the necessary set of experimental data one applies to the real plant (Σ) (with a zero initial condition) the inputs (6.21), with (6.22). For each $u_i(t)$, $i = \overline{1, m}$, one acquires the first $2(n+m)$ values of the augmented state vector $\tilde{x}$ (the values of $x_e(t)$ are gained by computation according to (6.16)).

For the numerical relevance of the considered samples, T must be chosen so that the $2(n+m)$ samples are acquired during the transient state.

Now one builds the matrix $\tilde{D} \in \mathbf{R}^{(n+m)\times 2(n+m)m}$ and extracts the matrices $\tilde{\mathcal{D}}_j \in \mathbf{R}^{(n+m)\times(n+m)}$, $j = \overline{0, n+m}$, in the same manner as in the single-input single-output case (see section 6.1).

Define the polynomial:

$$\tilde{P}(z) = \prod_{i=1}^{m} P_i(z)$$

with

$$P_i(z) = \sum_{j=0}^{\tilde{n}_i} \beta_{ij} z^j, \; \beta_{i\tilde{n}_i} = 1, \; i = \overline{1, m},$$

where $\tilde{P}(z)$ is associated with the set of zeroes $\tilde{Z}$ and $\tilde{n}_i$ denote the structural indices of the augmented discrete-time behavioural system (1.5), (6.16). According to Corollary 5.3 and Proposition 6.3 one can state the following result.

Theorem 6.3

In terms of Proposition 6.3 and Corollary 5.3, $\tilde{F}$ is given by:

$$\tilde{F} = \tilde{M}\tilde{K},$$

where $\tilde{M}$ denotes the β -parameter matrix of the augmented system, and

$$\tilde{K}^{\tau} \triangleq [\tilde{k}_1, \ldots, \tilde{k}_m] \in \mathbf{R}^{(n+m)\times m}$$

is built with the unique solutions of the following linear algebraic systems:

$$\begin{bmatrix} \tilde{\mathcal{R}}_{n+m}^{\tau} & 0 \\ \sum_{j=0}^{\tilde{n}_i} \beta_{ij} \tilde{\mathcal{D}}_j^{\tau} & \tilde{\mathcal{D}}_0^{\tau} \end{bmatrix} \begin{bmatrix} \tilde{g}_i \\ \tilde{k}_i \end{bmatrix} = [0 \ldots 0 \underbrace{1}_{n+i} 0 \ldots 0]^{\tau}, \; i = \overline{1, m},$$

where $\tilde{g}_i \in R^{n+m}$, $i = \overline{1,m}$, are some auxiliary vectors. ■

A self-learning strategy for determining the structurally stable compensator and representing the solution of problem $(\mathbf{P}')$ follows (*mutatis mutandis*) the procedure presented in section 6.2.

Part IV.

Linear Quadratic Optimal Control based on Non-Recursive Behavioural Models

CHAPTER 7

OPEN-LOOP OPTIMAL CONTROL

7.1. Problem statement

Consider the real plant (Σ) described by an extended linear non-recursive behavioural model $(X_{2N},\ldots,X_N,U_N)$, with $U_N = \alpha\Theta$, Θ being orthogonal (see Definition 2.2), and $N \geq \mu$ (see §3.2.b and equality (3.5)), with a sampling and holding period $T > 0$ satisfying Theorem 3.1. According to (2.6) for $j = N$ and respectively to (2.12) for $j = n$ and $1 \leq k \leq n$, one can write for any extended linear non-recursive behavioural model:

$$R_N = \alpha^{-2} X_N U_N^{\tau}, \qquad (7.1)$$

$$\Phi_k = E^k = \mathcal{X}_{n+k} \mathcal{X}_n^{-1}, \; 0 \leq k \leq n, \qquad (7.2)$$

where $\mathcal{X}_{n+k}, k = \overline{0,n}$, are built from $X_{n+k}, k = \overline{0,n}$, respectively, as it has been shown at the beginning of section 5.2.

Using the piecewise constant control (2.1), compactly expressed by (2.3), one can describe any evolution of the real plant (Σ), for any initial condition $x_0 = x(0)$, by the following non-recursive behavioural equation:

$$x_j = \Phi_j x_0 + R_j \overline{u}_j, \; j = \overline{0,N}, \text{ with } R_0 \overline{u}_0 \triangleq 0, \qquad (7.3)$$

where R_j can be obtained from (7.1) ($R_j, j = \overline{1,N}$, is constructed from the first j submatrices of the dimension $n \times m$ of R_N) and Φ_j is given by (7.2).

Now the point is how to use the information contained in any linear extended non-recursive behavioural model of the real plant (Σ) (allowing one to obtain (7.1)–(7.3)) in order to elaborate the piecewise optimal control which transfers the system state from any finite and fixed initial state $x_0 \in \mathbf{R}^n$ to any finite (fixed or free) final state $x_f \in \mathbf{R}^n$ along a finite and fixed time horizon $t_f = NT$, and which minimizes the functional:

$$J = (1/2)x_N^\tau M x_N + (T/2)\sum\nolimits_{j=0}^{N-1}(x_j^\tau Q x_j + u_j^\tau P u_j), \quad (7.4)$$

where M, Q and P are real symmetric matrices, having adequate dimensions, with M, Q being positive semi-definite and P positive definite and $x_f = x_N$.

The answer is suggested by the developments in Chapters 3 and 4 and it will be formulated in the sequel. For this purpose, let us denote:

$$x_{0j} \triangleq E^j x_0 = \Phi_j x_0,\ j = \overline{0, N}, \quad (7.5)$$

$$R_{0j} \triangleq \begin{cases} [0_{n\times(N-j)m}, R_j] \in \mathbf{R}^{n\times Nm},\ j = \overline{1, N-1}, \\ R_{00} \triangleq 0_{n\times Nm}, j = 0;\ R_{0N} \triangleq R_N, j = N \end{cases} \quad (7.6)$$

$$Q_j \triangleq \begin{cases} Q, & j = \overline{0, N-1} \\ MT^{-1}, & j = N \end{cases}, \quad (7.7)$$

$$S \triangleq \mathrm{diag}\{\underbrace{P, \ldots, P}_{N}\} + \sum\nolimits_{j=0}^{N} R_{0j}^\tau Q_j R_{0j} \in \mathbf{R}^{Nm\times Nm}, \quad (7.8)$$

$$p_0 \triangleq \sum\nolimits_{j=0}^{N} R_{0j}^\tau Q_j x_{0j} \in \mathbf{R}^{Nm}, \quad (7.9)$$

$$\rho_0^2 \triangleq \sum\nolimits_{j=0}^{N} x_{0j}^\tau Q_j x_{0j} \in \mathbf{R}_+. \quad (7.10)$$

Using (7.5) and (7.6) the system evolution described by equation (7.3) can be expressed as follows:

$$x_j = x_{0j} + R_{0j}\overline{u}_N, \ j = \overline{0,N}. \tag{7.11}$$

Substituting appropriately (7.11) into (7.4) and taking into account (7.7)–(7.10), one can successively write:

$$J=(1/2)(x_{0N}+R_{0N}\overline{u}_N)^\tau M(x_{0N}+R_{0N}\overline{u}_N)+$$

$$+(T/2)\sum_{j=0}^{N-1}(x_{0j}+R_{0j}\overline{u}_N)^\tau Q(x_{0j}+R_{0j}\overline{u}_N)+(T/2)\sum_{j=0}^{N-1}u_j^\tau P u_j=$$

$$=(T/2)\overline{u}_N^\tau \operatorname{diag}\{\underbrace{P,\dots,P}_{N}\}\overline{u}_N+(T/2)\overline{u}_N^\tau\left(\sum_{j=0}^{N-1}R_{0j}^\tau Q R_{0j}\right)\overline{u}_N+$$

$$+(T/2)\overline{u}_N^\tau R_{0N}^\tau M T^{-1} R_{0N}\overline{u}_{0N}+T\left(\sum_{j=0}^{N-1}x_{0j}^\tau Q R_{0j}\right)\overline{u}_N+$$

$$+T x_{0N}^\tau M T^{-1} R_{0N}\overline{u}_N+(T/2)\sum_{j=0}^{N-1}x_{0j}^\tau Q x_{0j}+(T/2)x_{0N}^\tau M T^{-1}x_{0N},$$

and finally:

$$J = (T/2)(\overline{u}_N^\tau S\overline{u}_N + 2p_0^\tau \overline{u}_N + \rho_0^2). \tag{7.12}$$

7.2. Case of a free final state

Theorem 7.1

For any finite and fixed time horizon t_f and for any sampling and holding period $T>0$ (according to Theorem 3.1) satisfying $t_f = NT, N \geq \mu$, the piecewise constant control (2.1) which transfers the system state from $x_0 \in \mathbf{R}^n$ to any finite and free final state $x_f = x_N \in \mathbf{R}^n$ and minimizes the functional (7.4), with (7.5)–(7.10), is uniquely expressed by:

$$\overline{u}_N^* = -S^{-1}p_0, \tag{7.13}$$

and the actual final state is given by (7.11) for $j=N$ and $\overline{u}_N=\overline{u}_N^*$.

Proof: As for $u(t)$ and $x(t)$, no constraints have been previously stated, and the necessary condition, so that J has an extremum, is the following:

$$\frac{\partial J}{\partial \overline{u}_N} = 0. \tag{7.14}$$

Applying (7.14) for (7.12), one obtains:

$$S\overline{u}_N^* = -p_0. \tag{7.15}$$

Since

$$\frac{\partial^2 J}{\partial \overline{u}_N^2} = TS$$

and S is obviously a positive definite symmetric matrix (see (7.8) where $\operatorname{diag}\{P,...,P\}$ is positive definite and the second term is a sum constituted of positive semi-definite matrix terms), it follows that J has an absolute minimum and (7.15) has the solution (7.13). The optimal trajectory is:

$$x_k^* = \Phi_k x_0 - R_{0k} S^{-1} p_0,\ k = \overline{0, N}, \tag{7.16}$$

$$x_0^* = x_0, \quad x_N^* = x_N,$$

and J achieves for $\overline{u}_N^*$ its absolute minimum:

$$J^* = (T/2)(\rho_0^2 - p_0^\tau S^{-1} p_0). \qquad \blacksquare$$

Remark 7.1

The hypothesis that U_N in (7.1) is orthogonal is not necessary and it can be relaxed to det $U_N \neq 0$ as it has been specified in Definition 2.3. The orthogonality of U_N is advantageous for obtaining R_N using (7.1). ■

Remark 7.2

In the case of some pre-specified constraints for $u(t)$ and/or $z(t)$ (e.g., amplitude constraints) one must apply the method of quadratic programming or other suitable ones depending on the analytical formulation of the constraints. ■

Remark 7.3

The functional (7.4) may be generalized in a certain sense, for instance as follows:

$$J = (1/2)e_N^\tau M e_N + (T/2)\sum_{j=0}^{N-1}(e_j^\tau Q e_j + u_j^\tau P u_j),$$

where

$$e_j = x_{dj} - x_j,\ j = \overline{0, N},$$

is the error between a desired trajectory $x_{dj}, j = \overline{0, N},$ and the trajectory $x_j, j = \overline{0, N},$ performed by the system. In this case, Theorem 7.1 is to be applied *mutatis mutandis*. ■

7.3. Case of a fixed final state

The final state x_N being previously given, the first term in J is constant (and also known) and consequently it is a redundant term which may be avoided (e.g., by taking $M = 0$). It follows that, for the next developments, (7.8)–(7.10) become:

$$S \triangleq \text{diag}\{\underbrace{P, \ldots, P}_{N}\} + \sum_{j=0}^{N-1} R_{0j}^{\tau} Q R_{0j}, \tag{7.17}$$

$$p_0 \triangleq \sum_{j=0}^{N-1} R_{0j}^{\tau} Q x_{0j}, \tag{7.18}$$

$$\rho_0^2 \triangleq \sum_{j=0}^{N-1} x_{0j}^{\tau} Q x_{0j}. \tag{7.19}$$

Theorem 7.2

For any finite and fixed time horizon t_f and for any sampling and holding period $T > 0$ (according to Theorem 3.1) satisfying $t_f = NT$, $N \geq \mu$, the piecewise constant control (2.1) which transfers the system state described by (7.11), from $x_0 \in \mathbf{R}^n$ to any finite and fixed final state $x_f = x_N \in \mathbf{R}^n$ and minimizes the functional (7.4) with (7.5)–(7.7), (7.17)–(7.19), is uniquely expressed by:

$$\overline{u}_N^* = S^{-1}[R_N^{\tau}(R_N S^{-1} R_N^{\tau})^{-1}(R_N S^{-1} p_0 + x_N - x_{0N}) - p_0]. \tag{7.20}$$

Proof: One has already shown in section 7.1 that (7.4) is equivalent to (7.12), but in this case S, p_0 and ρ_0 are defined by (7.17), (7.18) and (7.19), respectively. The problem to be dealt with consists in the minimization of the functional (7.12) under the condition:

$$R_N \overline{u}_N = x_N - x_{0N}, \tag{7.21}$$

which results from (7.3) and (7.5) for $j = N$ and for the given x_N and x_{0N}. This is equivalent to the minimization of the following extended functional:

$$J_e = J + \psi^{\tau}(R_N \overline{u}_N - x_N + x_{0N}), \tag{7.22}$$

where $\psi \in \mathbf{R}^n$ is the vector of Lagrange multipliers, because for $\overline{u}_N = \overline{u}_N$ (i.e., satisfying (7.21), too) it results in:

$$J_e^* = J^*.$$

The necessary condition so that J_e has an extremum is:

$$\frac{\partial J_e}{\partial \overline{u}_N} = 0, \qquad (7.23)$$

and applying (7.23) for (7.22) with (7.12) and (7.17)–(7.19), one obtains:

$$S\overline{u}_N = -p_0 - T^{-1}R_N^{\tau}\psi. \qquad (7.24)$$

Since

$$\frac{\partial^2 J_e}{\partial \overline{u}_N^2} = TS$$

and S is a positive definite symmetric matrix, it follows that J_e has an absolute minimum and (7.24) has the solution:

$$\overline{u}_N^* = -S^{-1}(p_0 + T^{-1}R_N^*\psi). \qquad (7.25)$$

Substituting (7.25) into (7.21) results in:

$$\psi = -T(R_N S^{-1} R_N^{\tau})^{-1}(R_N S^{-1} p_0 + x_N - x_{0N}),$$

which together with (7.25) yields (7.20). The optimal trajectory is uniquely expressed by:

$$\begin{aligned} x_k^* = \Phi^k x_0 + R_{0k} S^{-1}[R_N^{\tau}(R_N S^{-1} R_N^{\tau})^{-1}(R_N S^{-1} p_0 + \\ + x_N - x_{0N}) - p_0], \qquad (7.26) \\ k = \overline{0, N},\ x_0^* = x_0,\ x_N^* = x_N. \ \blacksquare \end{aligned}$$

It must be underlined here that Remarks 7.1–7.3 are also valid, *mutatis mutandis*, for the case of the fixed final state.

7.4. Minimal energy control

This is a special case of the functional (7.4) in which $M = 0$, $Q = 0$ and $P = I_m$, so that (7.4) has the following form:

$$J = (T/2)\sum_{j=0}^{N-1}\left\|u_j\right\|_2^2 = (T/2)\left\|\overline{u}_N\right\|_2^2 \tag{7.27}$$

as it has already been stated by (3.8) with $t_f = NT$ and (2.1), (2.3).

The problem of minimal energy control will occasion the investigation (in a simple manner) of certain aspects regarding the effects of decreasing the sampling and holding period $T > 0$ upon the minimal value of the polynomial when the final state x_N is given on the unit hypersphere in $\mathbf{R}^n$.

To simplify the next developments, let us assume that $x_0 = 0$ and x_N is fixed, so that S, p_0 and ρ_0^2 in (7.17)–(7.19) become: $S = I_{Sm}$, $p_0 = 0$ and $\rho = 0$. Under these circumstances, from Theorem 7.2 one may easily derive the following result.

Theorem 7.3

For any finite and fixed time horizon t_f and for any sampling and holding period $T > 0$ (according to Theorem 3.1) satisfying $t_f = NT$, $N \geq \mu$, the piecewise constant control (2.1) which transfers the system state, described by (7.11), from $x_0 = 0$ to any finite and fixed final state $x_f = x_N \in \mathbf{R}^n$ and minimizes the energy functional (7.27) is uniquely expressed by:

$$\overline{u}_N^* = R_N^+ x_N, \tag{7.28}$$

where $R_N^+ = R_N^\tau (R_N R_N^\tau)^{-1}$ is the (unique) Moore–Penrose pseudo-inverse of R_N. ■

Remark 7.4

Solution (7.28) is the minimal norm solution of the linear algebraic system obtained from (7.3) for $j = N, x_0 = 0$ and pre-specified x_N. In this respect, using (7.1), $\overline{u}_N^*$ is the minimal norm solution of the following linear algebraic system:

$$\alpha^{-2} X_N U_N^\tau \overline{u}_N^* = x_N .$$

The numerical solution of this system is advantageous by using the SVD of matrix X_N. One first solves the under-determined system:

$$X_N \tilde{\overline{u}}_N^* = \alpha^2 x_N$$

and then:

$$U_N^\tau \overline{u}_N^* = \tilde{\overline{u}}_N^* .$$ ■

Remark 7.5

According to section 3.2.b, by determining the controllability index $n_\delta(T)$ of the discrete-time behavioural model for the given sampling and holding period $T > 0$, one can transfer the state from the origin to any finite and fixed final state with the minimum number of time steps $N = n_\delta(T) \le \mu$, for the finite and "free" time horizon $t_f = NT$. Obviously, for the class of controls defined by (2.1), and depending on $x_N \in \mathbf{R}^n$, this may represent the minimum time control with pre-specified T. This conclusion, which is transparent from (7.28) requiring $N \ge n_\delta(T)$ for the existence of the Moore–Penrose pseudoinverse R_N^+, is also true in terms of Theorem 7.2. Thus, in a certain sense, Theorems 7.2 and 7.3 also include the case of finite and "free" final time $t_f = NT$ with pre-specified T and $N = n_\delta(T)$. More details about the minimal-time control problem will be given in section 8.3. ■

Johnson (1969) introduces an infimax index having an energy significance for the evaluation of controllability quality:

$$I = \max_{\|x\|_2=1} \min_{u(t)} J \qquad (7.29)$$

where J is the functional defined by (3.8). It is shown that:

$$I = 1/(2\lambda_{\min}) \qquad (7.30)$$

where $\lambda_{\min}$ denotes the minimum eigenvalue of the Gram controllability matrix:

$$\int_0^{t_f} e^{AT} BB^{\tau} e^{A^{\tau}t} dt$$

associated with the continuous-time linear recursive behavioural model (1.4).

The evaluation of I given by (7.29) over the class of controls (2.1), for $t_f = NT$, $N \in \mathbf{N}$, $N \geq \mu$, leads to:

$$I_T = \max_{\|x_N\|_2=1} \min_{\bar{u}_N \in \mathbf{R}^{Nm}} (T/2) \|\bar{u}_N\|_2^2 . \qquad (7.31)$$

Using solution (7.28) one obtains:

$$\min_{\bar{u}_N \in \mathbf{R}^{Nm}} \|\bar{u}_N\|_2^2 = x_N^{\tau} (R_N R_N^{\tau})^{-1} x_N ,$$

so that (7.31) becomes:

$$I_T = (T/2) \max_{\|x_N\|_2=1} [x_N^{\tau} (R_N R_N^{\tau})^{-1} x_N] = (T/2)(1/r_n^2), \qquad (7.32)$$

where r_n denotes the minimum singular value of R_N (see Theorem 3.3).

Now, in terms of these constructions, one can prove the following result.

Theorem 7.4

Consider system (1.4) controlled by (2.1) on a finite and fixed time horizon t_f, with the sampling and holding periods $T{=}t_f/N$ and $T'{=}t_f/\nu N$, $N \geq \mu$, $\nu \in \mathbf{N}$, $\nu \geq 2$ (for T and T' except for the values specified by Theorem 3.1). Then the following inequality holds:

$$I_{T'} \leq I_T. \tag{7.33}$$

Proof: According to the proof of Theorem 4.2, denoting by r_n, r_n' the minimum singular values of $R_N, R_{\nu N}$ respectively, the following inequality holds:

$$r_n \leq r_n'.$$

This result together with (7.32) yields (7.33). ■

Corollary 7.1

The decreasing sampling and holding period T implies the decreasing infimax energy index I_T and for T approaching zero (i.e., N approaching infinity) I_T approaches I defined by (7.29).

Proof: From Proposition 7 proved in Moore (1981) one can easily obtain:

$$\lim_{N\to\infty} I_T = I$$

and the monotonicity of the sub-sequences according to Theorem 7.4 proves the corollary completely. ■

Taking into account the relations between the minimum singular values r_n and s_n of matrices R_N and X_N respectively, once again one can evidence the usefulness of the linear non-recursive behavioural models for the approximation of the infimax energy index I by the index I_T for a sufficiently small sampling and holding period T.

Theorem 7.5

Consider system (1.4) controlled on a finite and fixed time horizon t_f by (2.1), with the sampling and holding period $T = t_f / N$, $N \geq \mu$ (except for the values of T according to Theorem 3.1). Then for any linear non-recursive behavioural model (X_N, U_N) with orthogonal test signals, the following equality holds:

$$I_T = \frac{\alpha^2 T}{2} \frac{1}{s_n^2},$$

where s_n denotes the minimum singular value of matrix X_N.

Proof: This results from (7.32) and $s_n = \alpha r_n$ (see the proof of Theorem 3.2). ■

CHAPTER 8

CLOSED-LOOP OPTIMAL CONTROL

In this chapter, the problem of state feedback optimal control based on the knowledge of an extended linear non-recursive behavioural model of the real plant (Σ) is addressed, using the results in sections 7.2 and 7.3.

It is well known that the solution to this problem, relying on the knowledge of the discrete-time linear recursive behavioural model (1.5), consists in using the linear state feedback controller:

$$u(t) \triangleq u_k = F(k)\, x(kT),\ t \in [kT,(k+1)T),\ k \in \mathbf{N}_0. \quad (8.1)$$

To minimize the functional (7.4) with a finite and free final state, $F(k)$ is computed as follows:

$$F(N-k) = -[P + H^{\tau}\Omega(k-1)H]^{-1} H^{\tau}\Omega(k-1)E, \quad k=\overline{1, N}, \quad (8.2)$$

and $\Omega(k)$ is the solution to the following discrete Riccati matrix equation:

$$\begin{aligned}\Omega(k) &= [E + HF(N-k)]^{\tau}\,\Omega(k-1)[E + HF(N-k)] + \\ &\quad + F^{\tau}(N-k)PF(N-k) + Q,\ k=\overline{1, N},\end{aligned} \quad (8.3)$$

with

$$\Omega(0) = MT^{-1}. \quad (8.4)$$

8.1. Case of a free final state

Our concern is the adequate use of the open-loop optimal control (7.13) and he optimal state trajectory (7.16) in order to obtain an equation whose solution is $F(k)$.

For the sake of simplicity, let us denote by $[\Delta]_{(i)}, i=\overline{1,N}$, the row blocks of dimension $m \times p$ composing the matrix $\Delta \in \mathbf{R}^{Nm \times p}$.

Theorem 8.1

For any finite and fixed time horizon $t_f > 0$ and for any sampling and holding period $T > 0$ (according to Theorem 3.1) satisfying $t_f = NT$, $N \geq \mu$, the matrices $F(k), k=\overline{0,N-1}$, of the time-variant controller (8.1) which transfers the system state described by (7.11), from $x_0 \in \mathbf{R}^n$ to any finite and free final state $x_f = x_N \in \mathbf{R}^n$ and minimizes the functional (7.4) with (7.5)–(7.10), are solutions to the following equations:

$$F(k)(\Phi_k - R_{0k} S^{-1} \sum_{j=0}^{N} R_{0j}^{\tau} Q_j \Phi_j) = -[S^{-1}]_{(N-k)} \sum_{j=0}^{N} R_{0j}^{\tau} Q_j \Phi_j, \quad k=\overline{0,N-1}, \tag{8.5}$$

provided that they are consistent.

Proof: Substituting (7.5) into (7.9) results in:

$$p_0 = \sum_{j=0}^{N} R_{0j}^{\tau} Q_j E^j x_0 \tag{8.6}$$

and further, by inserting (8.6) into (7.13) and (7.16), one obtains:

$$\overline{u}_N^{\tau} = -S^{-1} \sum_{j=0}^{N} R_{0j}^{\tau} Q_j E^j x_0, \tag{8.7}$$

$$x_k^* = (E^k - R_{0k} S^{-1} \sum_{j=0}^{N} R_{0j}^{\tau} Q_j E^j) x_0, \; k=\overline{0,N}. \tag{8.8}$$

Now, extracting u_k^* from (8.7), substituting it, together with (8.8) in (8.1) and taking into account that x_0 is an arbitrary vector, in the end (8.5) is obtained. ■

The consistency conditions of equations (8.5) may be easily avoided if one intends to use an optimal control recursive law instead of (8.1). This possibility is proved by the following result.

Theorem 8.2

In terms of Theorem 7.1, optimal control (7.13) can be expressed by the following formula:

$$u_k^* = -[S^{-1}]_{(N-k)} \sum_{j=0}^{N} R_{0j}^{\tau} Q_j \Phi_j \Phi_k^{-1}(x_k^* - R_k \bar{u}_k^*),\ k = \overline{0, N-1}, \quad (8.9)$$

where $R_0 \bar{u}_0^* \triangleq 0$.

Proof: Since

$$\det E = \det e^{AT} = e^{\operatorname{Tr}(A)T} \neq 0, \quad (8.10)$$

one can also use (7.3), for $j = k$ in order to express the backward solutions:

$$x_0 = E^{-k}(x_k^* - R_k \bar{u}_k^*),\ k = \overline{0, N}. \quad (8.11)$$

Now substituting (8.11) into (8.7) results in:

$$\bar{u}_N^* = -S^{-1} \sum_{j=0}^{N} R_{0j}^{\tau} Q_j E^{j-k}(x_N^* - R_k \bar{u}_k^*),\ k = \overline{0, N}. \quad (8.12)$$

Finally, extracting the vector u_k^* from (8.12) one obtains (8.9). ■

8.2. Case of a fixed final state

Consider the open-loop optimal control (7.20) and the optimal state trajectory (7.26). Taking into account that x_N has been given previously, now one must find a state feedback controller described by the following equation:

$$u(t) \triangleq u_k = G(k)x_N + F(k)x(kT),\ t\in[kT,(k+1)T),\ k\in \mathbf{N}_0, \quad (8.13)$$

where $G(k)$ is an appropriate set point matrix and $F(k)$ has the same significance as in (8.1).

Theorem 8.3

Consider an arbitrary finite and fixed time horizon $t_f > 0$ and a sampling and holding period $T > 0$ (according to Theorem 3.1) satisfying $t_f = NT$, $N \geq \mu$. Matrices $F(k)$ and $G(k)$, $k = \overline{0, N-1}$, of the time-variant controller (8.13) which transfers the system state, described by (7.11), from $x_0 \in \mathbf{R}^n$ to any finite and fixed final state $x_f \in \mathbf{R}^n$ and minimizes the functional (7.4), with (7.5), (7.6), (7.17)–(7.19), are solutions of the following equations:

$$F(k)(\Phi_k - R_{0k}S^{-1}L) = -[S^{-1}]_{(N-k)}L,\ k = \overline{0, N-1}, \quad (8.14$$

$$G(k) = \{[S^{-1}]_{(N-k)} - F(k)R_{0k}S^{-1}\}D\ k = \overline{0, N-1}, \quad (8.15)$$

where:

$$D \triangleq R_N^\tau (R_N S^{-1} R_N^\tau)^{-1} \in \mathbf{R}^{Nm\times n}, \quad (8.16)$$

$$L \triangleq [D\Phi_N + (I_{Nm} - DR_N S^{-1})\sum_{j=0}^{N-1} R_{0j}^\tau Q \Phi_j] \in \mathbf{R}^{Nm\times n}, \quad (8.17)$$

provided that equations (8.14) are consistent.

Proof: Substituting (7.5) into (7.18) results in:

$$p_0 = \sum_{j=0}^{N-1} R_{0j}^\tau Q E^j x_0 \quad (8.18)$$

and replacing (8.18) into (7.20) and (7.26), together with (8.16) and (8.17), one can write:

$$\overline{u}_N^* = -S^{-1}Lx_{0+} + S^{-1}Dx_N, \quad (8.19)$$

$$x_k^* = (E^k - R_{0k}S^{-1}L)x_0 + R_{0k}S^{-1}Dx_N,\ k = \overline{0,N}. \quad (8.20)$$

Now, extracting u_k^* from (8.19), replacing this result and (8.20) into (8.13) and considering that x_0 and x_N are arbitrary vectors, one obtains (8.14) and (8.15). ■

As in the case of a free final state one can derive a recursive formula for (7.20).

Theorem 8.4

In terms of Theorem 7.2, optimal control (7.20) can be expressed by the following formula:

$$u_k^* = -[S^{-1}]_{(N-k)}L\Phi_k^{-1}(x_k^* - R_k\overline{u}_k^*) + [S^{-1}]_{(N-k)}Dx_N,\ k = \overline{0,N-1}, \quad (8.21)$$

where D and L are defined by (8.16) and (8.17).

Proof: According to (8.10) one can use (8.11) in order to replace x_0 in (8.19). This results in:

$$\overline{u}_N^* = -S^{-1}LE^{-k}(x_k^* - R_k\overline{u}_k^*) + S^{-1}Dx_N,\ k = \overline{0,N}, \quad (8.22)$$

and by extracting the vector u_k^* from (8.22), one obtains (8.21). ■

8.3. Optimal non-recursive behavioural models

Consistent with our non-recursive behavioural model-based approach, it is natural to take into consideration the possibility to build optimal non-recursive behavioural models in order to synthesize controllers (8.1) and (8.3) directly from experimental data.

Unlike the definitions of the non-recursive behavioural models given in Chapter 2 and according to the purpose above formulated, one builds the optimal non-recursive behavioural models by applying a set of n different open-loop optimal control inputs to a real plant (Σ) and by acquiring the data representing the optimal state evolutions accordingly. Thus, one will have in view the following matrices:

$$U_k^* = [u_k^{*1},...,u_k^{*n}] \in \mathbf{R}^{m\times n},\ k=\overline{0,N-1}, \tag{8.23}$$

and

$$X_k^* = [x_k^{*1},...,x_k^{*n}] \in \mathbf{R}^{n\times n},\ k=\overline{0,N-1}, \tag{8.24}$$

which, under certain conditions (to be detailed below), can be used for obtaining, on an experimental basis, the matrices defining (8.1) and (8.13), respectively.

a. Free final state

For generating a set of n optimal control vectors of the form (8.7), which obviously depend on the initial conditions, one chooses a set of n linearly independent initial conditions $x_0^{*1},...,x_0^{*n}$. Under these circumstances, matrices U_k^* and $X_k^*, k=\overline{0,N-1}$, can be completely known on an experimental basis, and, according to (8.1), one may state the following result.

Theorem 8.5

Under the hypotheses of Theorem 8.1, matrices $F(k), k=\overline{0,N-1}$, are solutions of the following equations:

$$F(k)X_k^* = U_k^*,\ k=\overline{0,N-1}, \tag{8.25}$$

provided that they are consistent.

Proof: It is easy to see that by using (8.7) and (8.8), one can write:

$$U_k^* = -[S^{-1}]_{(N-k)}\sum_{j=0}^{N} R_{0j}^{\tau} Q_j E^j X_0^*,\ k=\overline{0,N-1}, \tag{8.26}$$

$$X_k^* = (E^k - R_{0k}S^{-1}\sum_{j=0}^{N} R_{0j}^{\tau} Q_j E^j)X_0^{\tau},\ k=\overline{0,N-1}. \tag{8.27}$$

For $\det X_k^* \neq 0$, multiplying (8.5) to the right by X_0^* and taking into account (8.26) and (8.27), result in (8.25). Hence Theorems 8.1 and 8.5 are proved to be equivalent. ■

b. Fixed final state

Now one must determine the pairs of matrices $F(k), G(k)$, $k = \overline{0, N-1}$, in (8.13). In order to first obtain $F(k)$ and then $G(k)$ the experiments should be organized according to the following two distinct steps:

1° One takes $x_N = 0$ and generates a set of n optimal control vectors of the form (8.19) corresponding to a set of n linearly independent initial conditions $x_{01}^{*1}, \ldots, x_{01}^{*n}$ respectively. By experimental measurement one builds the matrices U_{k1}^* and $X_{k1}^*, k = \overline{0, N-1}$.

2° One takes x_0 and generates a set of n optimal control vectors of the form (8.19) corresponding to a set of n linearly independent final conditions $x_{N_2}^{*1}, \ldots, x_{N_2}^{*n}$ respectively. By experimental measurement one builds the matrices U_{k2}^* and $X_{k2}^*, k = \overline{0, N-1}$.

Thus, according to (8.13), one can formulate the following result.

Theorem 8.6

Under the hypotheses of Theorem 8.3 matrices $F(k)$ and $G(k)$ $k = \overline{0, N-1}$ are solutions of the following equations:

$$F(k) X_{k1}^* = U_{k1}^*, \ k = \overline{0, N-1}, \tag{8.28}$$

$$G(k) X_{N2}^* = U_{k2}^* - F(k) X_{k2}^*, \ k = \overline{0, N-1}, \tag{8.29}$$

provided that equations (8.28) are consistent.

Proof: Using (8.19) and (8.20) and considering the two-step procedure for building $U_{k1}^*, X_{k1}^*, k=\overline{0,N-1},$ and $U_{k2}^*, X_{k2}^*, k=\overline{0,N-1},$ one can write:

$$U_{k1}^* = -[S^{-1}]_{(N-k)} L X_{01}^*,\ k=\overline{0,N-1}, \tag{8.30}$$

$$X_{k1}^* = (E^k - R_{0k} S^{-1} L) X_{01}^*,\ k=\overline{0,N-1}, \tag{8.31}$$

$$U_{k2}^* = -[S^{-1}]_{(N-k)} D X_{N2}^*,\ k=\overline{0,N-1}, \tag{8.32}$$

$$X_{k2}^* = R_{0k} S^{-1} D X_{N2}^*,\ k=\overline{0,N-1}. \tag{8.33}$$

For $\det X_{01}^* \neq 0$ by multiplying (8.14) to the right by X_{01}^* and considering (8.31) and (8.32), one obtains (8.28). Finally, for $\det X_{N2}^* \neq 0$ by multiplying (8.15) to the right by X_{N2}^* and considering (8.33) and (8.34), one obtains (8.29). As a matter of fact, Theorems 8.3 and 8.6 are equivalent. ■

8.4. Minimal-time state feedback control

Based on the results given in Wonham (1979), showing that the state subspace of system (1.5) which can be transferred to the origin in no more than $j \geq 0$ time steps is given by $Im(Z_j)$, where:

$$Z_j = [E^{-1}H, \ldots, E^{-j}H],\ j=\overline{1,n}, \tag{8.34}$$

one can synthesize the minimal-time state feedback control using the information existing in any linear non-recursive behavioural model of the real system (Σ) as discussed in Voicu and Pastravanu (1990). A solution based on the linear recursive behavioural model (1.5) was given in Leden (1977) in order to design a dead-beat control.

Theorem 8.7

For any linear extended non-recursive behavioural model $X_{2n}, \ldots, X_n, U_n)$ with U_n orthogonal and with a sampling and holding period $T > 0$ according to Theorem 3.1 associated with the real plant (Σ) a state feedback control (1.6) which ensures for the closed-loop system the minimal time transfer from any initial state $x_0 \in \mathbf{R}^n$ to the origin is the unique solution of the matrix equation:

$$\begin{bmatrix} \mathcal{R}_N^\tau & 0 \\ \mathcal{X}_{2n}^\tau & \mathcal{X}_n^\tau \end{bmatrix} \begin{bmatrix} G^\tau \\ F^\tau \end{bmatrix} = \begin{bmatrix} 0^\tau_{(n-m)\times m} & I_m & 0^\tau_{n\times m} \end{bmatrix}^\tau . \tag{8.35}$$

Matrices $\mathcal{X}_n^\tau, \mathcal{X}_{2n}^\tau$ and $\mathcal{R}_N^\tau$ have been defined in 5.2, and G is an auxiliary unknown matrix.

Proof: For a given $G \in \mathbf{R}^n$ the second group of equations of (8.35) yields:

$$F = -G\mathcal{X}_{2n}\mathcal{X}_n^{-1}. \tag{8.36}$$

According to the relation:

$$\mathcal{X}_{2n} = E^n \mathcal{X}_n$$

(see (2.12)), from (2.37) one can obtain:

$$F = -GE^n. \tag{8.37}$$

On the other hand, a dead-beat state feedback controller L is given by the unique solution of the matrix equation (as in Leden (1977)):

$$LW = [0_{m\times(n-m)} \vdots -I_m] \tag{8.38}$$

with

$$W = [E^{-n_\delta} H^{(n_\delta)}, \ldots, E^{-2}H^{2)}, E^{-1}H^{(1)}] \in \mathbf{R}^{n\times n},$$

where n_δ is the controllability index of the discrete-time recursive behavioural model (1.5) and the notations $E^{-k}H^{(k)}, 1 \le k \le n_\delta$, are used in Leden (1977) to indicate the linearly independent columns progressively selected from the matrices $E^{-1}H, E^{-2}H, ..., E^{-k}H$.

Since

$$\mathcal{R}_n = E^n W,$$

which relies on the equality:

$$R_n = E^n[E^{-n}H, E^{-n+1}H, \ldots, E^{-1}H],$$

it results that equation (8.38) can be rewritten as follows:

$$(LE^{-n})\mathcal{R}_n = [0_{m\times(n-m)}, -I_m],$$

In other words, the unique solution of the first group of equations of (8.35), i.e.,

$$G\mathcal{R}_n = [0_{m\times(n-m)}, I_m]$$

has the following expression:

$$G = -LE^n,$$

which together with (8.37) leads to $F = L$.

Thus, we have shown that each initial state of the real system (Σ) can be transferred to the origin in the minimal time. ■

Remark 8.1

Equation (8.35) can be obtained as a special case of equations (5.14) from Theorem 5.3 for $P_i(z) = z^n$, $i = \overline{1,m}$, i.e., for $P(z) = z^n$. Consequently, as it is well known, the dead-beat control (8.1) with F given by (8.35) assigns all the eigenvalues in $z = 0$ and the closed-loop system matrix is nilpotent of index n^δ as used in Leden (1977). ■

Remark 8.2

Since:

$$(E + FH)ImR_i \subset ImR_{i-1},\ i = \overline{1,k},\ 1 \le k \le n_\delta,$$

$$ImR_0 = 0,$$

it follows that if $x_0 \in ImR_k$ and $x_0 \notin ImR_{k-1}$ the dead-beat state feedback controller transfers the system state to the origin in precisely k time steps.■

PART V.

EXPERIMENTS AND CONCLUSIONS

CHAPTER 9

SIMULATION EXPERIMENTS

This chapter illustrates, by simulation experiments, the theoretical approach based on non-recursive behavioural models in order to solve problems of state controllability analysis and structural and optimal synthesis.

The considered examples refer to systems studied in several papers devoted to state controllability analysis and structural and optimal synthesis, which rely on the knowledge of recursive behavioural models. In the light of our approach, the recursive behavioural models of the considered examples have been used only for generating, by numerical simulation, the necessary collection of data, which allows one to construct non-recursive behavioural models. The numerical simulation of recursive behavioural models has been performed using MATLAB; the same software has been used for solving different problems of linear algebra requested by the implementation of the proposed algorithms.

9.1. Controllability analysis for single-input systems

Example 9.1

Consider the following linear recursive behavioural model:

$$\begin{bmatrix} \dot{x}_1 \\ \vdots \\ \dot{x}_8 \end{bmatrix} = \begin{bmatrix} -1/2^0 & & 0 \\ & \ddots & \\ 0 & & -1/2^7 \end{bmatrix} \begin{bmatrix} x_1 \\ \vdots \\ x_8 \end{bmatrix} + \begin{bmatrix} 1 \\ \vdots \\ 1 \end{bmatrix} u,$$

which has been constructed in the manner suggested by Paige (1981) in order to obtain a numerically ill-conditioned state controllability matrix.

For different orthogonal matrices Θ, automatically generated by the procedure presented in section 2.2, one has correspondingly generated (by numerical simulation) different linear non-recursive behavioural models $(\tilde{\tilde{X}}_8, U_8)$ where $\tilde{\tilde{X}}_8$ has the significance considered in section 3.3. For the singular values of $\tilde{\tilde{X}}_8$ one has obtained $\tilde{\tilde{s}}_8 / \tilde{\tilde{s}}_1 = 1.3151 \cdot 10^{-8}$ for $T = 2$ sec and $\tilde{\tilde{s}}_8 / \tilde{\tilde{s}}_1 = 7.373 \cdot 10^{-7}$ for $T = 4$ sec. For all Θ considered in these experiments, the first four significant decimal digits of the results obtained for $T = 2$ sec and $T = 4$ sec, respectively, have been identical.

The numerical simulation of the linear recursive behavioural model (performed in order to generate the non-recursive one) has been performed in double precision, involving MEAC $= 10^{-10}$ and COAC = $16^{-13} \times P(n)$, where $P(n)$ denotes a small degree polynomial depending on $\rho = n$. Since AC < $\tilde{\tilde{s}}_8 / \tilde{\tilde{s}}_1$ for both $T = 2$ sec and $T = 4$ sec, then, in accordance with (3.6), one can conclude that the system is completely state-controllable.

It is obvious that the small value of MEAC has allowed us to illustrate the usefulness of non-recursive behavioural models in a limited case which involves numerical difficulties for the controllability analysis based on linear recursive behavioural models.

From the practical point of view, this example is quite artificial because real measurement cannot be performed with MEAC = $8 \cdot 10^{-10}$. For practical testing, where MEAC usually has larger values, the message of the routine pointing out the rank defectiveness for X_n represents an essential criterion in analysing the (real) practical state controllability. AC may often represent a severe constraint for the threshold because of the norm inequalities used in section 3.3. For the considered system, under the usual precision of the data acquisition equipment (12-16 bit) one may expect to ascertain that for the corresponding, ratio $\tilde{\tilde{s}}_i / \tilde{\tilde{s}}_1, i = \overline{1,7}$, the system order is 5 or 6. This reflects the great discrepancy among the energy values required to control different state variables; this discrepancy is also emphasized by the diagonal canonical form of the linear recursive behavioural model. In the case of a real application, one uses an equivalent reduced order behavioural model, preferably based on the pole dominance concept (Litz, 1979).

Example 9.2

This is built in the same manner, suggested in Paige (1981), in order to investigate the geometry of reachability sets corresponding to the energy or amplitude constraints imposed on the control input. In this respect the following linear recursive behavioural model is considered:

$$\begin{bmatrix} \dot{x}_1 \\ \dot{x}_2 \\ \dot{x}_3 \end{bmatrix} = \begin{bmatrix} -1.00 & 0 & 0 \\ 0 & -0.50 & 0 \\ 0 & 0 & -0.25 \end{bmatrix} \begin{bmatrix} x_1 \\ x_2 \\ x_3 \end{bmatrix} + \begin{bmatrix} 1 \\ 1 \\ 1 \end{bmatrix} u,$$

which has the advantage of intuitive interpretations in $\mathbf{R}^3$.

The state controllability Gram matrix for the finite horizon t_f has been computed using the facilities of symbolic manipulation software and has the following form:

$$G(t_f) = \begin{bmatrix} \frac{1}{2}(1-e^{-2t_f}) & \frac{2}{3}(1-e^{-\frac{3}{2}t_f}) & \frac{4}{5}(1-e^{-\frac{5}{4}t_f}) \\ \frac{2}{3}(1-e^{-\frac{3}{2}t_f}) & (1-e^{-t_f}) & \frac{4}{3}(1-e^{-\frac{3}{4}t_f}) \\ \frac{4}{5}(1-e^{-\frac{5}{4}t_f}) & \frac{4}{3}(1-e^{-\frac{3}{4}t_f}) & 2(1-e^{-\frac{1}{2}t_f}) \end{bmatrix}.$$

For $t_f = 3$ sec it results:

$$G(3) = \begin{bmatrix} 0.49876062391167 & 0.65926066897451 & 0.78118580331519 \\ 0.65926066897451 & 0.95021293163214 & 1.19280103391751 \\ 0.78118580331519 & 1.19280103391751 & 1.55373967970314 \end{bmatrix}$$

Determining the eigenvalues and the eigenvectors of $G(3)$, the result is that for piecewise continuous input signals, constrained by (4.1), the reachability set is an ellipsoid (E) which has the directions of the axes given by the vectors:

$$\begin{bmatrix} 0.38952354222118 \\ 0.57005835740206 \\ 0.72339814708881 \end{bmatrix}, \begin{bmatrix} -0.80219533434243 \\ -0.17589595582280 \\ 0.57056398263860 \end{bmatrix}, \begin{bmatrix} 0.45249757525837 \\ -0.80255472204782 \\ 0.38878253883637 \end{bmatrix}$$

and the lengths of the semi-axes:

$$1.70714410044898, 0.29613284337498, 0.02603064777995.$$

For the same horizon $t_f = 3$ sec, using linear non-recursive behavioural models (generated by numerical simulation with the help of the recursive behavioural model) for different sampling and holding periods T_i, the reachability sets are as follows:

1. Ellipsoid (E_1) for $T_1 = 1$ sec, which has the directions of the axes given by the vectors:

$$\begin{bmatrix} 0.38138839253809 \\ 0.56881048904391 \\ 0.72869576751264 \end{bmatrix}, \begin{bmatrix} -0.79378530308330 \\ -0.20247591806758 \\ 0.57350535761372 \end{bmatrix}, \begin{bmatrix} 0.47375920745264 \\ -0.79715627712282 \\ 0.37429144152321 \end{bmatrix}.$$

and the lengths of the semi-axes:

$$1.69346930363781,\ 0.26353915612110,\ 0.01666073755965.$$

2. Ellipsoid (E_2) for $T_2 = 0.5$ sec, which has the directions of the axes given by the vectors:

$$\begin{bmatrix} 0.38738151039025 \\ 0.56975898865165 \\ 0.72478290560582 \end{bmatrix}, \begin{bmatrix} -0.80007737940204 \\ -0.18283738451529 \\ 0.57135512406276 \end{bmatrix}, \begin{bmatrix} 0.45805212844930 \\ -0.80121481868114 \\ 0.38502345636183 \end{bmatrix}$$

and the lengths of the semi-axes:

$$1.70362271539822, 0.28766069732059, 0.02363231890760.$$

3. Ellipsoid (E_3) for $T_3 = 0.1$ sec which has the directions of the axes given by the vectors:

$$\begin{bmatrix} 0.38943635819905 \\ 0.57004657101146 \\ 0.72345437298474 \end{bmatrix}, \begin{bmatrix} -0.80211047592435 \\ -0.17617764097907 \\ 0.57059637505986 \end{bmatrix}, \begin{bmatrix} 0.45272299182289 \\ -0.80250130572923 \\ 0.38863034747406 \end{bmatrix}$$

and the lengths of the semi-axes:

$$1.70700183830462, 0.29578938073888, 0.02593361601509.$$

It is obvious that by using linear non-recursive behavioural models and decreasing the sampling and holding period T one can obtain reachability sets which approximate, better and better, the reachability set associated with the state controllability Gram matrix.

The differences which occur among the state variables x_1, x_2 and x_3 from the controllability quality point of view are practically revealed by the shape of the ellipsoids $(E_i), i = \overline{1,3}$, and can be investigated without any knowledge of the mathematic behavioural model of the considered system.

Similar geometric characterizations can be obtained using piecewise constant control inputs that satisfy amplitude constraints of the form (4.10).

For instance, the reachability polyhedron (P_1) corresponding to $T_1 = 1$ sec has the vertices given by the following matrix:

columns 1 through 4

0.950212	0.779116	0.485124	−0.3140281
1.553739	0.974742	0.599134	−0.0201376
2.110533	1.037220	0.732372	0.3409400,

columns 5 through 8

0.314028	−0.779116	−0.485124	−0.9502129
0.020137	−0.974742	−0.599134	−1.5537396
−0.340940	−1.037220	−0.732374	−2.1105337

and is depicted in Figure 9.1.

Finally, it must be mentioned that this simulation experiment allowed us to use accurate linear non-recursive behavioural models even in the case of small sampling and holding periods.

In real-life applications, the accuracy of non-recursive behavioural models obviously depends on the measurement precision and word length of the available A/D converters.

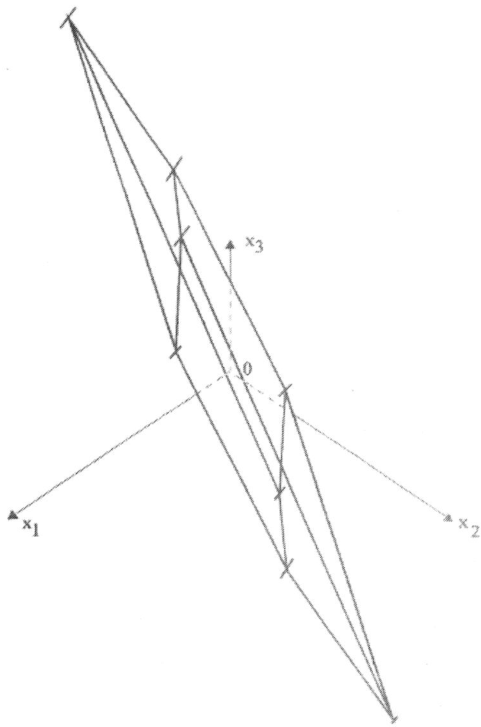

Figure 9.1: Reachability polyhedron (Example 9.2) for $T = 1\text{sec}$ and $t_f = 3\text{sec}$.

9.2. Structural synthesis for single-input systems

Example 9.3

Consider the internally ill-balanced system given in Moore (1981):

$$\begin{bmatrix} \dot{x}_1 \\ \dot{x}_2 \\ \dot{x}_3 \\ \dot{x}_4 \end{bmatrix} = \begin{bmatrix} 0 & 0 & 0 & -150 \\ 1 & 0 & 0 & -245 \\ 0 & 1 & 0 & -113 \\ 0 & 0 & 1 & -19 \end{bmatrix} \begin{bmatrix} x_1 \\ x_2 \\ x_3 \\ x_4 \end{bmatrix} + \begin{bmatrix} 4 \\ 1 \\ 0 \\ 0 \end{bmatrix} u$$

$$y = \begin{bmatrix} 0 & 0 & 0 & 1 \end{bmatrix} \begin{bmatrix} x_1 & x_2 & x_3 & x_4 \end{bmatrix}^{\tau} .$$

For a sampling and holding period $T = 0.05$ sec, using the linear non-recursive behavioural models obtained by numerical simulation of the continuous-time linear recursive behavioural model, the results are:

$$\text{cod}_1 S(4 \times 0.050) = 3.738 \cdot 10^{-13}$$

and

$$\text{cod}_2 S(4 \times 0.050) = 6.114 \cdot 10^{-7},$$

which confirm that the system is internally ill-balanced.

Relying on the unitary step response obtained by numerical simulation, one intends to synthesize a state feedback compensator which incorporates an internal behavioural model for the exogenous step, by applying the results given in Chapter 6.

Matrix $D \in \mathbf{R}^{5\times10}$ built with the procedure presented in section 6.1, has the following form:

columns 1 through 4:

0.199966276687	0.199565178297	0.198397427661	0.196326030709
0.054944573043	0.064280310506	0.072325695583	0.078814299591
0.001307361115	0.003990116008	0.006531758716	0.008756443204
0.000017338262	0.000098964405	0.000215165912	0.000336504281
0.000000000000	–0.000017338262	–0.000116302667	–0.000331468579,

columns 5 through 8:

0.193374526792	0.189642800173	0.185260120685	0.180359942164
0.083739518813	0.087222790403	0.089439418110	0.090578181227
0.010603146167	0.012071634502	0.013191455627	0.014004530320
0.000448244126	0.000543957104	0.000621765313	0.000682160208
–0.000667972860	–0.001116216986	–0.001660174091	–0.002281939404,

columns 9 through 10:

0.175067245313	0.169492970232
0.090820515291	0.090330834158
0.014555710330	0.014887941013
0.000726757110	0.000757601359
–0.002964099613	–0.003690856724.

An overshot of about 5% and a settling time of about 4 seconds are required for the closed-loop system. These performances correspond to the following pair of dominant poles:

$$s_{1,2} = -0.7 \mp j0.7.$$

Applying Theorem 6.1, one wishes to assign the following eigenvalues:

$$\begin{aligned} z_{1,2} &= e^{s_{1,2}T} = 0.965 \mp j0.0338 \\ z_3 &= 0.7788 \\ z_4 &= 0.7408 \\ z_5 &= 0.7047, \end{aligned}$$

which are the roots of the polynomial:

$$\tilde{P}(z)=z^5-4.1543z^4+6.8731z^3-5.6606z^2+2.32102z-0.37906.$$

Solving equation (6.11) with $\alpha=1$, one obtains:

• auxiliary vector:

$$\tilde{g}=[0.105743\ \text{-}\ 0.422964\ \ 1.690811\ \text{-}\ 6.689396\ \text{-}\ 0.063423]^{\tau}10^{6}$$

• feedback gain vector:

$$\tilde{f}=[-0.017495\ \ 0.063222\ \ -0.342745\ \ 2.489809\ \ 0.025422]^{\tau}10^{2}.$$

The numerical simulation of the unitary step response of the closed-loop system built with $\tilde{f}$ is represented in Figure 9.2 and confirms the achievement of the imposed performances.

One must comment here on the large value of $\|\tilde{g}\|_2=6.913824\cdot10^6$ which in fact represents $1/(\mathrm{Pr}_p E^4 h)$. As a natural consequence, $\|\tilde{f}\|_2$ has a relatively large value, too.

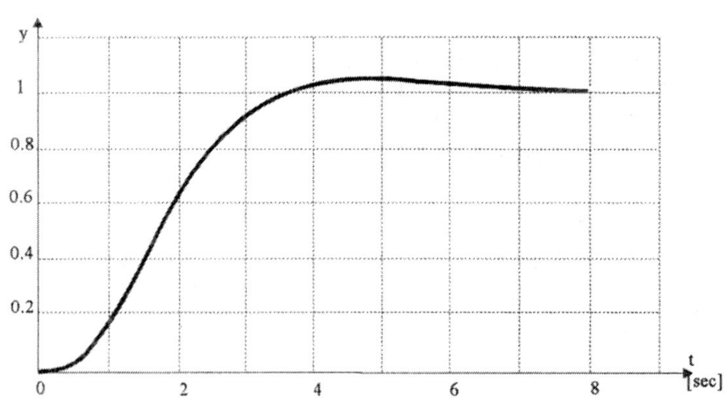

Figure 9.2: Unit step response of the closed-loop system (Example 9.2)

The utilization of the sampling and holding period $T = 0.05$ sec is not possible in real-time applications when 12-bit A/D converters are used for data acquisition, because the resulting linear non-recursive behavioural model cannot be sufficiently accurate (see Remark 3.1). Actually, the value of $\mathrm{cod}_2 S(4\times 0.050)$ given above, indicates that the practical usage of the state controllability of this system requires precision, MEAC = 10^{-8}– 10^{-7} in data acquisition (see Chapter 3) which can be easily ob-tained in the case of simulation experiments. In the case of real data acquisition based on the usual A/D converters, the resulting linear non-recursive behavioural models should indicate an incomplete state controllability for the considered system.

9.3. Controllability analysis and structural synthesis for multi-input systems

Example 9.4

Consider the system given in Leden (1977):

$$\dot{x} = Ax + Bu,\ x \in \mathbf{R}^7,\ u \in \mathbf{R}^2,$$

with:

$$\begin{bmatrix} -a_1 & a_2 & & & & & 0 \\ b_2 & -b_1 & b_2 & -b_3 & & & \\ -b_3 & b_2 & -b_1 & b_2 & -b_3 & & \\ & -b_3 & b_2 & -b_1 & b_2 & -b_3 & \\ & & -b_3 & b_2 & -b_1 & b_2 & -b_3 \\ & & & -b_3 & b_2 & -b_1 & b_2 \\ 0 & & & & & a_2 & -a_1 \end{bmatrix},$$

$$\begin{bmatrix} a_2 & 0 \\ -b_3 & 0 \\ 0 & 0 \\ 0 & 0 \\ 0 & -b_3 \\ 0 & a_2 \end{bmatrix}, \qquad \begin{aligned} a_1 &= 0.732962\cdot 10^{-1} \\ a_2 &= 0.366301\cdot 10^{-1} \\ b_1 &= 0.916113\cdot 10^{-1} \\ b_2 &= 0.488402\cdot 10^{-1} \\ b_3 &= 0.305251\cdot 10^{-1}, \end{aligned}$$

which describes the heat diffusion process in a copper rod with a constant cross-section; u_1 and u_2 denote the controlled end temperatures, and x_i, $i=\overline{1,7}$, denote the temperatures in seven equidistant points of the rod. The process operates within the temperature range 20° to 30°C, all the variables in the continuous-time linear recursive behavioural model being referred to 25°C, conventionally considered as the origin of the operating range.

Via a numerical simulation of the step responses of the linear recursive behavioural model for several values of the sampling and holding period T, one has built linear extended non-recursive behavioural models by applying the procedure presented in 2.7.

The computation of $\text{cod}_1 S(7\times T_i)$ and $\text{cod}_2 S(7\times T_i)$ yielded:

1. For $T_1 = 60\,\text{sec}$ $\text{cod}_1 S(7\times T_1) = 6.5025\cdot 10^{-12}$ and $\text{cod}_2 S(7\times T_1) =$ $= 2.55\cdot 10^{-6}$;

2. For $T_2 = 30\,\text{sec}$ $\text{cod}_1 S(7\times T_2) = 19.7136\cdot 10^{-8}$ and $\text{cod}_2 S(7\times T_2) =$ $= 4.44\cdot 10^{-4}$;

3. For $T_3 = 15\,\text{sec}$ $\text{cod}_1 S(7\times T_3) = 1.87489\cdot 10^{-5}$ and $\text{cod}_2 S(7\times T_3) =$ $= 0.433\cdot 10^{-2}$.

In Leden (1977) the model coefficients have been given with 7 decimal digits, so that $\text{MEAC} = 7\sqrt{2}\cdot 10^{-7}$ has been considered. The result is

$$\text{AC} < \text{cod}_2 S(4 \times T_i) \text{ for all } T_i, i = \overline{1,3}.$$

From the physical point of view, taking into account that the largest time constant is $T_{\max} = 176\,\text{sec}$, according to Leden (1977), the above results confirm the fact that one can easily obtain the desired profile of temperatures in the rod by using the same number of time steps and decreasing the sampling and holding period. It is also obvious that by using smaller sampling and holding periods one can hardly obtain the desired profile of temperatures and this is in accordance with Remark 3.1.

Using (3.4), the state controllability matrices for T_i, $i = \overline{1,3}$, have been obtained. By means of the SVD procedure one has computed the structural indices $n_1 = 4$ and $n_2 = 3$; the resulting controllability index is $n_\delta = 4$.

For the same T_i, $i = \overline{1,3}$, and $t_f = 420\,\text{sec}$, the reachability sets which correspond to the energy constraint (4.1) have been determined:

Lengths of the semi-axes:

$T_1 = 60$ sec	$T_2 = 30$ sec	$T_3 = 15$ sec
Hyper-ellipsoid (E_1)	Hyper-ellipsoid (E_2)	Hyper-ellipsoid (E_3)
0.10893241118042	0.11354749715865	0.11444065263547
0.09560077480950	0.10431186943475	0.10835533724165
0.04227858987839	0.05498301159000	0.05775268268898
0.01133021575525	0.02491235355347	0.03398417555727
0.00113353936716	0.00660269957757	0.01321566803558
0.00003619059731	0.00091242373223	0.00375903517739
0.00000027781833	0.00005276987075	0.00056264731073

It is easy to recognize the effect of decreasing T_i, $i = \overline{1,3}$, upon the reachability sets; for instance, the smallest semi-axis length is going to be multiplied by about 2500 when the sampling and holding period value decreases 4 times (from 60 sec to 15 sec).

For $T_1 = 60$ sec and $T_2 = 30$ sec, respectively, one has determined the open-loop control input which minimizes the energy functional (3.8) with $t_f = 420$ sec and for $x_0 = 0$ and $x_f = x(420) = [1 \ldots 1]^{\tau}$.

1. For $T_1 = 60$ sec the following sequence of 7 time steps has been obtained:

0.	0.	0.	0.	0.55393	1.10331	1.00019
0.	0.	0.	12.70169	−2.90726	1.13347	1.00019;

2. For $T_2 = 30$ sec the following sequence of 14 time steps has been obtained:

0.	0.	0.	0.	0.	0.	0.
0.	0.	0.	0.	0.	0.	0.

0.	0.	0.	0.	−6.52342	1.55264	0.99555
0.	0.	0	17.63110	4.37174	0.58853	1.00433.

One has to remark that, in both cases, the control sequences have non-zero entries only for the last 4 time steps. This corresponds to the controllability index $n_\delta = 4$. At the same time, some control values in the above sequences are relatively large, unlike the case of the infinite horizon (i.e., $t_f = \infty$) characterized by smaller values.

For $t_f = 420$ sec and for the same non-recursive behavioural models, one has evaluated the infimax energy indices (see 7.4):

1. For $T_1 = 60$ sec, $I_{60} = 6.478108437042134 \cdot 10^{12}$;

2. For $T_2 = 30$ sec, $I_{30} = 1.795551819564512 \cdot 10^{8}$;

3. For $T_3 = 15$ sec, $I_{15} = 1.579419550329927 \cdot 10^{6}$.

These values confirm the improvement of I_{T_i} when T_i decreases, in accordance with Theorem 7.4.

In Leden (1977) the efficiency of the dead-beat control in getting uniform temperatures in the rod has been illustrated.

In order to synthesize the dead-beat control by applying the procedure presented in section 8.4 one has selected from R_7 (for $T_1 = 60$ sec) the first 7 linearly independent columns, so the matrix $\mathcal{R}_7$ has the following form:

$\mathcal{R}_7$, columns 1 through 4:

0.628989907878	0.104989055819	0.047523686611	0.000987214180
0.338185426235	0.159833749882	0.081820008795	0.004529667722
0.153160811336	0.157234727552	0.096367578368	0.017964044492
0.057817762456	0.119051630729	0.091722065192	0.057817762456
0.017964044492	0.073731533287	0.073771149462	0.153160811336
0.004529667722	0.038129578521	0.049664747338	0.338185426235
0.000987214180	0.015189186188	0.024436875250	0.628989907878,

$\mathcal{R}_7$, columns 5 through 7:

0.015189186188	0.028245611959	0.024436875250
0.038129578521	0.050728916354	0.049664747338
0.073731533287	0.063658634854	0.073771149462
0.119051630729	0.065652666625	0.091722065192
0.157234727552	0.057699336094	0.096367578368
0.159833749882	0.042255486488	0.081820008795
0.104989055819	0.022167510896	0.047523686611.

The characteristic parameters as in Ionescu (1986) of the discrete-time linear recursive behavioural model are highlighted by the following equalities (which are correlated with the structural indices $n_1 = 4$, $n_2 = 3$, and the first group of equations (5.15) in Corollary 5.3):

$$[H, EH, E^2H, E^3H]\begin{bmatrix} 0.00000006215508 \\ -0.00000006144876 \\ -0.00061220876875 \\ 0.00056910528078 \\ 0.08985986873333 \\ -0.04807626046725 \\ -0.76997586163660 \\ 0.00000000000000 \end{bmatrix} + E^4 h_1 = 0,$$

$$[H, EH, E^2H]\begin{bmatrix} 0.00000024846163 \\ -0.00000024846163 \\ -0.00237411305986 \\ 0.00237411305971 \\ 0.27249930590662 \\ -0.27249930590380 \end{bmatrix} + E^3H\begin{bmatrix} -1.00000000000366 \\ 1.00000000000000 \end{bmatrix} = 0.$$

This yields the organization of the β-parameters (in accordance with matrix form (1.14), as used by Corollary 5-3):

$$M = \begin{bmatrix} 1 & -1 \\ 0 & 1 \end{bmatrix}.$$

One has also extracted the matrices $\mathcal{D}_0, \mathcal{D}_3$ and $\mathcal{D}_4$ (which are correlated with the last group of equations (5.15) in Corollary 5.3):

$\mathcal{D}_0$, columns 1 through 4:

0.628989907878	0.000987214180	0.104989055819	0.015189186188
0.338185426235	0.004529667722	0.159833749882	0.038129578521
0.153160811336	0.017964044492	0.157234727552	0.073731533287
0.057817762456	0.057817762456	0.119051630729	0.119051630729
0.017964044492	0.153160811336	0:073731533287	0.157234727552
0.004529667722	0.338185426235	0.038129578521	0.159833749882
0.000987214180	0.628989907878	0.015189186188	0.104989055819,

$\mathcal{D}_0$, columns 5 through 7:

0.047523686611	0.024436875250	0.028245611959
0.081820008795	0.049664747338	0.050728916354
0.096367578368	0.073771149462	0.063658634854
0.091722065192	0.091722065192	0.065652666625
0.073771149462	0.096367578368	0.057699336094
0.049664747338	0.081820008795	0.042255486488
0.024436875250	0.047523686611	0.022167510896;

$\mathcal{D}_3$, columns 1 through 4:

0.028245611959	0.022167510896	0.018708392684	0.017106902752
0.050728916354	0.042255486488	0.034171532296	0.031938838526
0.063658634854	0.057699336094	0.043956966334	0.042386687288
0.065652666625	0.065652666625	0.046723621211	0.046723621211
0.057699336094	0.063658634854	0.042386687288	0.043956966334
0.042255486488	0.050728916354	0.031938838526	0.034171532296
0.022167510896	0.028245611959	0.017106902752	0.018708392684,

$\mathcal{D}_3$, columns 5 through 7:

0.012947776492	0.012525795960	0.009115452674
0.023806065898	0.023217767287	0.016801953449
0.030916423051	0.030502665535	0.021898859592
0.033236808654	0.033236808654	0.023642088690
0.030502665535	0.030916423051	0.021789837495
0.023217767287	0.023806065898	0.016646941047
0.012525795960	0.012947776492	0.009004263880;

$\mathcal{D}_4$, columns 1 through 4:

0.018708392684	0.017106902752	0.012947776492	0.012525795960
0.034171532296	0.031938838526	0.023806065898	0.023217767287
0.043956966334	0.042386687288	0.030916423051	0.030502665535
0.046723621211	0.046723621211	0.033236808654	0.033236808654
0.042386687288	0.043956966334	0.030502665535	0.030916423051
0.031938838526	0.034171532296	0.023217767287	0.023806065898
0.017106902752	0.018708392684	0.012525795960	0.012947776492,

$\mathcal{D}_4$, columns 5 through 7:

0.009115452674	0.009004263880	0.006459103880
0.016801953449	0.016646941047	0.011916855048
0.021898859592	0.021789837495	0.015552689601
0.023642088690	0.023642088690	0.016817097722
0.021789837495	0.021898859592	0.015523963072
0.016646941047	0.016801953449	0.011876010409
0.009004263880	0.009115452674	0.006429806442.

These allow us to apply Corollary 5.3 (see also Theorem 5.3), for $\alpha_{14}=1, \alpha_{1j}=0, j=\overline{0,3}$, and $\alpha_{23}=1, \alpha_{2j}=0, j=\overline{0,2}$, respectively. The resulting matrix K has the form (in accordance with (5.13)):

K, columns 1 through 4:

−0.468751386377	−0.716549604979	−0.978983682614	−1.064939266303
0.023396671369	0.012991045788	−0.036828229024	−0.135272025848,

K, columns 5 through 7:

−0.468751386377	−0.716549604979	−0.978983682614	−1.064939266303
0.023396671369	0.012991045788	−0.036828229024	−0.135272025848,

Consequently, the matrix $F = MK$, has the following entries:

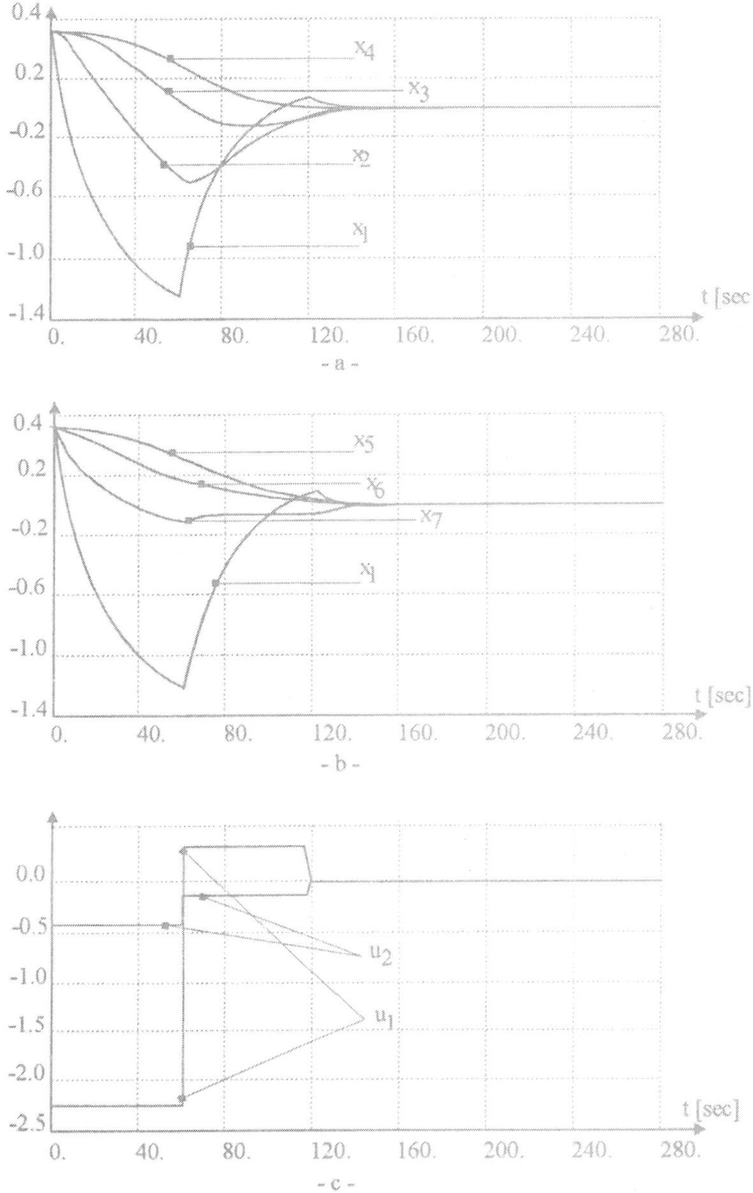

Figure 9.3: Numerical simulation of the rod-cooling process (Example 9.4)

F columns 1 through 4:

-0.492148057747	-0.729540650768	-0.942155453590	-0.929667240455
0.023396671369	0.012991045788	-0.036828229024	-0.135272025848,

F columns 5 through 7:

-0.741975999406	-0.458293646389	-0.245441208439
-0.237007683993	-0.258255957921	-0.223310178125.

The results of the numerical simulation of the rod-cooling process from 25.5°C to 25°C (conventionally from 0.5 to 0.0) are presented in Figure 9.3: a – the evolution of state variables x_1, x_2, x_3 and x_4, b – the evolution of state variables x_1, x_5, x_6 and x_7 and c – the evolution of dead-beat controls u_1 and u_2.

It can be seen that the cooling process is finished in about 120 sec (i.e., 2 time steps), and the control values are acceptable, i.e., the dead-beat control in this case can be an efficient control solution.

CHAPTER 10

REAL-TIME EXPERIMENTS FOR A CASE STUDY

10.1. Laboratory behavioural model

A scaled behavioural model of a two-coupled tank system, linearized around a given operating point, has been implemented on an analogue computer. The state vector $[x_1\ x_2]^T$ and the control input u represent the variations of the fluid level in the two reservoirs and the fluid flow, respectively; the output y is the fluid level in the first tank.

Figure 10.1 depicts the analogue behavioural model of the considered plant which corresponds to the following continuous-time linear recursive behavioural model:

$$\begin{bmatrix} \dot{x}_1 \\ \dot{x}_2 \end{bmatrix} = \begin{bmatrix} -0.5 & 0.5 \\ 0.25 & -0.25 \end{bmatrix} \begin{bmatrix} x_1 \\ x_2 \end{bmatrix} + \begin{bmatrix} 0.5 \\ 0 \end{bmatrix} u,$$

$$y = [1 \quad 0] \begin{bmatrix} x_1 \\ x_2 \end{bmatrix}.$$

The laboratory behavioural model has been interfaced by 12-bit A/D and D/A converters with an IBM compatible PC, running adequate real-time software.

10.2. Data acquisition and non-recursive behavioural model construction

The program devoted to data acquisition has the general structure presented in (Pastravanu, 1992) and allows one to save the acquired data corresponding to the step responses sampled with a given period T_0. A self-learning procedure supplies the non-recursive behavioural model for any value of the sampling and holding period T which is an integer multiple of T_0.

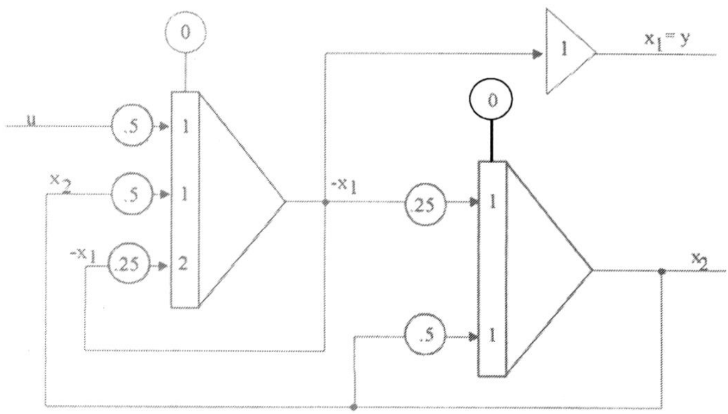

Figure 10.1: Analogue behavioural model of the two-coupled tank system

Thus, for our purposes, the continuous-time step responses plotted in Figure 10.2, which correspond to $u(t)=2\sigma(t)$, $u(t) = 2\sigma(t)$, have been sampled with $T_0 = 1\,\mathrm{sec}$.

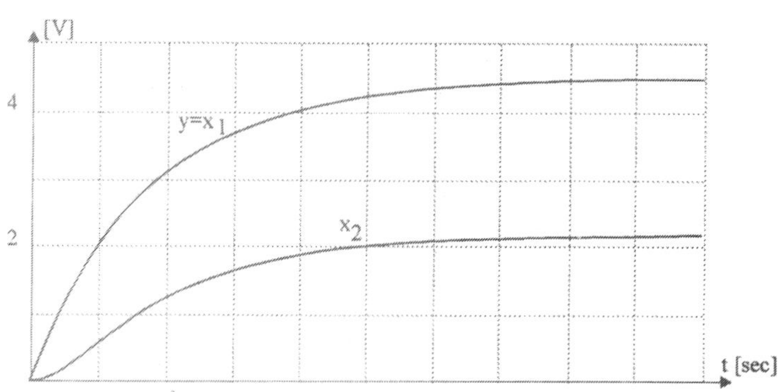

Figure 10.2: Step responses of the analogue behavioural model from Figure 10.1 for $u(t)=2\sigma(t)$.

10.3. Controllability analysis

a. Controllability degrees

Using the linear non-recursive behavioural model (i.e., finally matrix R_2) for four values of T, one determined $\mathrm{cod}_1 S(2 \times T_i)$, $\mathrm{cod}_2 (2 \times T_i)$ and $r_2(T_i)$ (in accordance with Theorem 3.3):

1. For $T_1 = 8\ \mathrm{sec}$:	2. For $T_2 = 4\ \mathrm{sec}$:
$\mathrm{cod}_1 S(2\times 8) = 0.0032$	$\mathrm{cod}_1 S(2\times 4) = 0.0110$
$\mathrm{cod}_2 S(2\times 8) = 0.0566$	$\mathrm{cod}_2 S(2\times 4) = 0.1048$
$r_2(8) = 0.0985;$	$r_2(4) = 0.1302;$
3. For $T_2 = 2\ \mathrm{sec}$:	4. For $T_2 = 1\ \mathrm{sec}$:
$\mathrm{cod}_1 S(2\times 2) = 0.0159$	$\mathrm{cod}_1 S(2\times 1) = 0.0084$
$\mathrm{cod}_2 S(2\times 2) = 0.1261$	$\mathrm{cod}_2 S(2\times 1) = 0.0917$
$r_2(2) = 0.1024;$	$r_2(1) = 0.0455.$

It is obvious that in the case of the piecewise constant control (2.1), from the controller point of view, the most advantageous value of T is within the range of 2-4 seconds; the measurement accuracy $\mathrm{MEAC} = 0.01$ should also be taken into account.

b. Reachability sets

Using the same linear non-recursive behavioural models, the reachability sets corresponding to energy and amplitude constraints, for a finite and fixed time horizon $t_f = 16\ \mathrm{sec}$, have been determined.

For the energy constraint (4.1) one has obtained the reachability sets depicted in Figure 10.3, i.e., the ellipses $E_i, i = \overline{1,4}$, for $T_1 = 8\ \mathrm{sec}$,

$T_2 = 4$ sec, $T_3 = 2$ sec, $T_4 = 1$ sec, respectively.

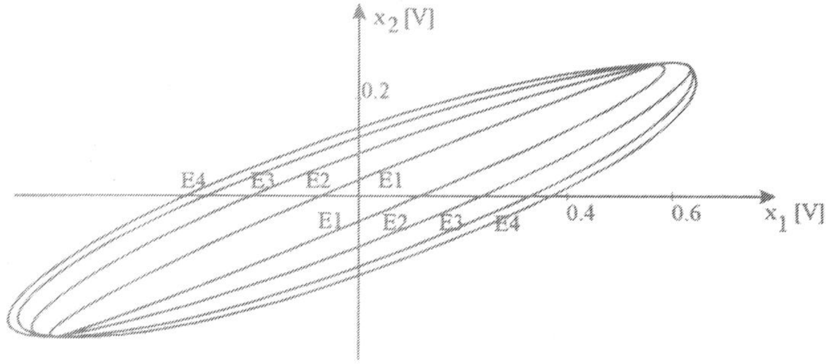

Figure 10.3: Reachability ellipses $E_i, i = \overline{1,4}$, of the system in Figure 10.1, for the energy constraint (4.1), with $t_f = 16$ sec, $T_1 = 8$ sec, $T_2 = 4$ sec, $T_3 = 2$ sec, $T_4 = 1$ sec. respectively

For the amplitude constraint (4.10), one has determined the reachability sets presented in Figure 10.4, i.e., the parallelograms $(P_i), i = \overline{1,2}$, for T_1=8 sec, T_2=4 sec, respectively.

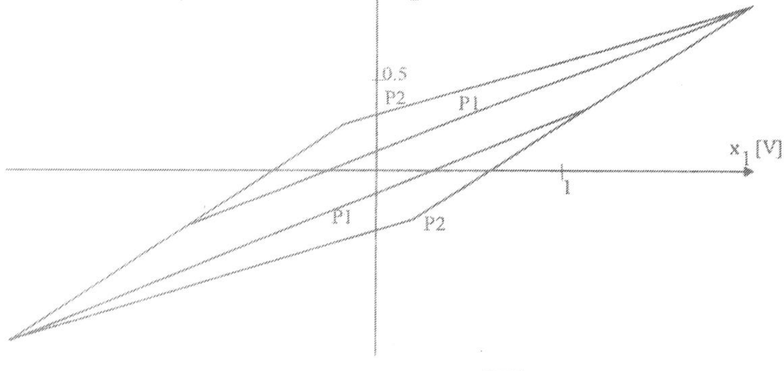

Figure 10.4: Reachability parallelograms $(P_i), i = \overline{1,2}$, for the amplitude constraint (4.10), with $t_f = 16$ sec, $T_1 = 8$ sec, $T_2 = 4$ sec, respectively

From Figures 10.3 and 10.4, one can easily see the effects of decreasing the sampling and holding period consisting in a perceptible enlargement of the reachability sets. At the same time, both representations highlight the system anisotropy which characterizes its forced response in $\mathbf{R}^3$.

10.4. Structural synthesis

The state feedback compensator which incorporates an internal behavioural model of the exogenous step has been designed using the linear non-recursive behavioural model for $T = 2$ sec and applying the procedure presented in section 6.1. The requested closed-loop performances are: an overshoot less than 10% and a stabilization time of about 10 sec. This may be achieved by assigning the following closed-loop eigenvalues:

$$z_{1,2} = 0.3792 \mp 0.3199,$$
$$z_3 = 0.1353,$$

which are the roots of the following polynomial:

$$\tilde{P}(z) = z^3 - 0.8937z^2 + 0.3487z - 0.0333$$

In accordance with the notations used in section 6.1, the augmented system is characterized by:

$$\tilde{D}_0 = \begin{bmatrix} 1.3620 & 0.7770 & 0.5370 \\ 10.2640 & 0.3950 & 0.3320 \\ 0.0000 & -1.3620 & -2.1390 \end{bmatrix},$$

$$\sum_{j=0}^{3} \beta_j \tilde{D}_j = \begin{bmatrix} 0.1456 & 0.1019 & 0.0853 \\ 0.0912 & 0.0711 & 0.0525 \\ -1.2388 & -1.3844 & -1.4863 \end{bmatrix},$$

which allowed one to determine the gain vector – see Theorem 6.1:

$$\tilde{f} = [-1.51955 \ \ 0.09816 \ \ 1.01749]^{\tau}.$$

A detailed analysis has shown that the matrix $\tilde{D}_0$ has a better numerical conditioning for $T = 2$ sec than for $T = 1$ sec and $T = 4$ sec a fact which once more justifies the adopted value $T = 2$ sec.

The state-controllability analysis of the augmented system (i.e., that incorporates the internal behavioural model of the exogenous step) revealed that the ratio of the extreme singular values of $\tilde{D}_0$ (in an exact expression 0.0282, due to $s_1(\tilde{D}_0) = 2.7860$, $s_3(\tilde{D}_0) = 0.0786$) ensures a reasonable numerical conditioning (in line with the discussions developed by section 5.3). At the same time, we are in the situation of satisfying a condition of type (5.17), with $n = 3$ and the measurement error $\varepsilon = 0.005$ (i.e., 5 mV). In accordance with Theorem 5.4, the relative errors that may appear when solving a system of form (5.11) are acceptable, for the considered laboratory experiment. We also see that the gain factors (collected by the matrix $\tilde{f}$) have convenient magnitudes, because the spectral norms $\left\|\tilde{D}_0^{-1}\right\|_2 = 12.72$ and $\|\sum_{j=0}^{3} \beta_j \tilde{D}_j\|_2 = 2.38$ are comparable in value.

Figure 10.5: Evolution of output y and control u for the closed-loop system; set point sequence: 0., 2., 0

Figure 10.5 depicts the evolutions of the output y and control u respectively, plotted simultaneously, for the set point sequence: $0., 2., 0$. It is obvious that the design specifications are achieved.

10.5. Open-loop minimal energy control

The most illustrative experiments have been realized by considering two final states which have the same norm and correspond (from the point of view of the anisotropy evidenced in paragraph 10.3.b) to a favourable direction in $\mathbf{R}^2$ and to an unfavourable one, respectively (meaning an orthogonal direction with respect to the first one – see the reachability ellipses from Figure 10.3).

In the first case, for the finite time horizon $t_f = 16\,\text{sec}$ and the final state $x_f = x^*(16) = [0.3\ \ 0.1]^\tau$, for $T_i, i = \overline{1,4}$, one obtained the following sequences for the minimal energy control:

1. For $T_1 = 8\,\text{sec}, N = 2$:	2. For $T_1 = 4\,\text{sec}, N = 4$:
$\overline{u}_2^* = [-0.1976\quad 0.2481]^\tau$;	$\overline{u}_4^* = [0.\ \ 0.\ \ 0.0501\ \ 0.2586]^\tau$;
3. For $T_1 = 2\,\text{sec}, N = 8$:	4. For $T_1 = 1\,\text{sec}, N = 16$:
$\overline{u}_8^* = [0.\ \ldots\ 0.3425\ \ 0.2451]^\tau$;	$\overline{u}_{16}^* = [0.\ \ldots\ 0.8489\ \ 0.\ \ 0.2952]^\tau$.

It should be emphasized that for $T_i, i = \overline{1,3}$, the minimal energy control is practically identical to that corresponding to $N = 2$, and $t_f = 2T_i$.

On the other hand, for $T_4 = 1\,\text{sec}$ and $N = 2$, i.e., $t_f = 2\,\text{sec}$ the following control sequence has been obtained:

$$\overline{u}_2 = [1.1785\ \ \text{-}\ 0.0842]^\tau$$

which corresponds to an energy value of 1.3960, which is larger than the value of 0.8078 associated with $\overline{u}_{16}^*$.

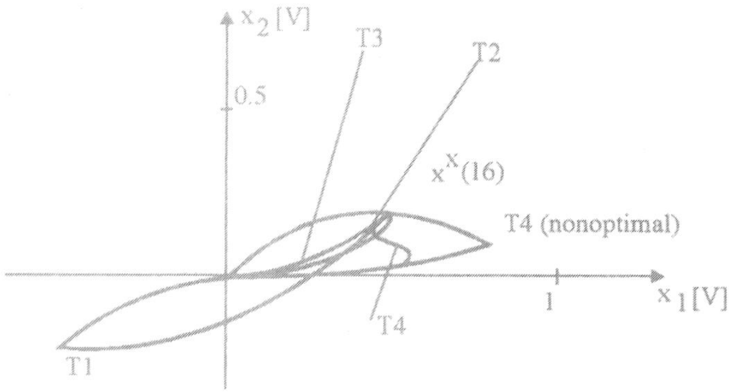

Figure 10.6: Minimal energy state space trajectories (for $T_i, i=\overline{1,4}$) and a non-optimal one (for T_4, $N=2$) for $x_f = x^*(16) = [0.3\ \ 0.1]^\tau$

Figure 10.6 represents the optimal state trajectories of the analogue behavioural model in Figure 10.1 for $T_i, i=\overline{1,4}$, and the non-optimal one for the situation $T_4 = 1$ sec and $N = 2$.

In the second case, for the same $t_f = 16$ sec and the final state $x_f = x^*(16) = [-0.1\ \ 0.3]^\tau$, for $T_i, i=\overline{1,4}$, one obtained the following sequences for the minimal energy control:

1. For $T_1 = 8$ sec, $N = 2$:	2. For $T_1 = 4$ sec, $N = 4$:
$\overline{u}_2^* = [3.0720\ \ 0.889]^\tau$;	$\overline{u}_4^* = [0.\ 0.\ 2.1885\ 1.0522]^\tau$;
3. For $T_1 = 2$ sec, $N = 8$:	4. For $T_1 = 1$ sec, $N = 16$:
$\overline{u}_8^* = [0. \ldots 0.\ 2.6137\ 1.6379]^\tau$;	$\overline{u}_{16}^* = [0. \ldots 0.\ 3.9509\ 0.\ -2.3855]^\tau$.

As in the first case, it can be remarked that for $T_i, i=\overline{1,3}$, the minimal energy control is practically identical to that corresponding to $N = 2$ and $t_f = 2T_i$. Unlike this, for T_4 and $N = 2$, i.e., $t_f = 2$ sec, the following control sequence has been obtained:

$$u_2 = [5.48477\ \ -4.1512]^\tau$$

which corresponds to an energy value of 47.3144, which is larger than the energy value of 21.3002 realized by $\overline{u}^*_{16}$.

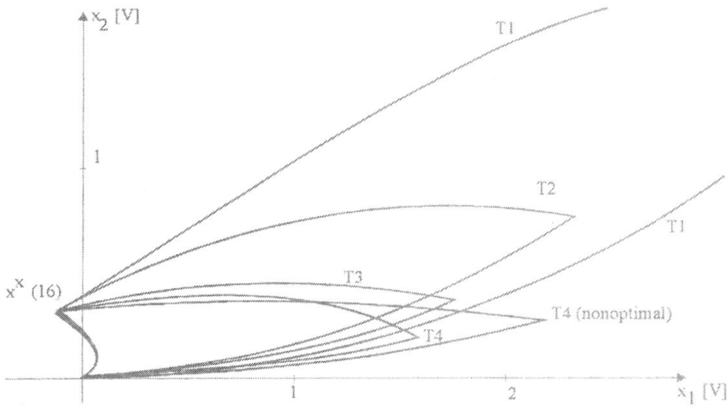

Figure 10.7: Minimal energy state space trajectories (for $T_i, i = \overline{1,4}$, and a non-optimal one (for T_4, $N = 2$) for $x_f = x^*(16) = [-0.1\ 0.3]^\tau$,

Comparing the experimental results of the two cases of open-loop optimal control and, especially, the trajectories plotted in Figure 10.6 and 10.7, respectively, one can confirm (as already expected) that in the second case the control of the plant requires more energy than in the first case.

10.6. Closed-loop optimal control

The system is to be transferred from $[0 \quad 0]^\tau$ (considered as the initial operating point) to the final state $[1 \ -0.5]^\tau$ to minimize the performance index (7.4) with $N = 2, T = 2\,\text{sec}, M = 0_{2\times 2}, P = 1$. Despite the simplicity, this case is particularly interesting because it illustrates the possibility of fulfilling the complementary evolutionary conditions required for reaching the desired final state: the level in the second tank should be decreased, while the level in the first one, should be increased. The experimental results are illustrated by Figure 10.8, where the initial time is 2 sec.

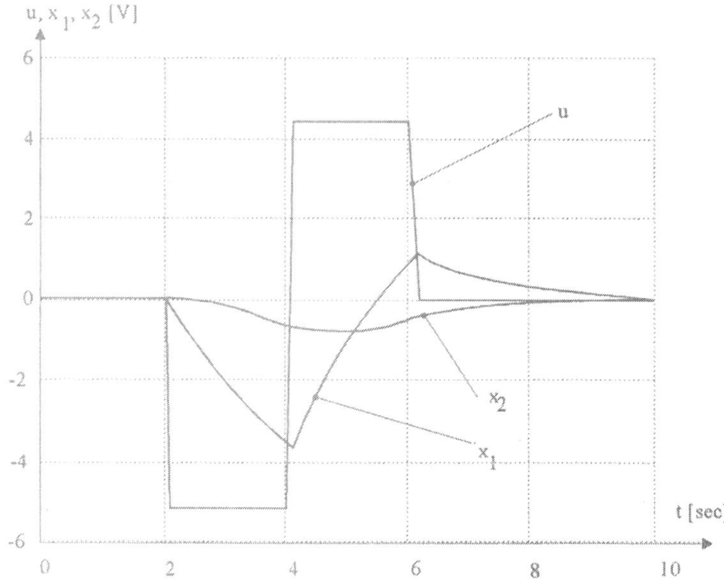

Figure 10.8: Control input and closed-loop response

The optimal non-recursive behavioural model built experimentally:

$$X_{01}^{*}=\begin{bmatrix}1.4111 & 0.7861\\ 0.2539 & 0.3906\end{bmatrix},\quad X_{11}^{*}=\begin{bmatrix}-0.5664 & -0.5908\\ 0.1611 & 0.1611\end{bmatrix},$$

$$U_{01}^{*}=[-1.9618\ \ -1.6435],\quad U_{11}^{*}=[0.\ \ 2.],$$

$$X_{02}^{*}=\begin{bmatrix}0. & 0.\\ 0. & 0.\end{bmatrix},\ X_{12}^{*}=\begin{bmatrix}0. & 1.4160\\ 0. & 0.2588\end{bmatrix},\ X_{22}^{*}=\begin{bmatrix}1.4169 & 0.8105\\ 0.2588 & 0.4150\end{bmatrix},$$

$$U_{02}^{*}=[0.\ \ 2.],\ U_{12}^{*}=[0.\ \ 2.],$$

yields the state feedback controller:

$$F(0)=[\text{-}0.9926\ \ \text{-}\,2;2095],\ F(1)=[\text{-}0.5213\ \ 0.1451],$$
$$G(0)=[\text{-}1.3695\ \ \ 7.4933],\ G(1)=[\ 1.7166\ \text{-}1.6645].$$

The resulting control input and the closed-loop response are plotted in Fig. 10.8, for: $F(k)=[0,\ 0.],\ G(k)=[0,\ 0.],\ k \geq 2$.

CHAPTER 11

FINAL OVERVIEW – CONCLUDING REMARKS

Starting from the geometric background of pole assignment which expresses a direct relation between the open-loop state trajectory and the state feedback matrix to ensure the assignment of a desired closed-loop spectrum (Chapter 1), the introduction of the non-recursive behavioural models (Chapter 2) becomes very natural. It offers the possibility to solve the synthesis problems (**P**) and (**P'**) which consist of the design of a state feedback discrete-time controller and a structurally stable discrete-time compensator, respectively. The procedures avoid the proper identification of a recursive behavioural model associated with the plant dynamics.

In fact, the non-recursive behavioural models which contain the same information about the system behaviour as the recursive behavioural model, allow both the controllability analysis (including its qualitative characterization based on experiments – Chapter 3) and the evaluation of the reachability sets (corresponding to energy or amplitude constraints, and quantized control input – Chapter 4).

Consistent with the geometric background of pole assignment it is shown that the non-recursive behavioural models allow the design of state feedback controllers (Chapter 5) and structurally stable compensators (Chapter 6), for both single- and multi-input cases. These new theoretical results which use direct, experimental knowledge of the system dynamics, that is fundamentally different from the classic framework, reformulate the solutions of the synthesis problems (**P**) and (**P'**) as easy-to-implement self-learning procedures.

Further, by relying on the knowledge of the non-recursive behavioural models, the linear quadratic optimal control is considered for a finite and fixed timc horizon and a finite (free or fixed) final state; the problem is first addressed for the open-loop case (Chapter 7) and then for the closed-loop case (Chapter 8), which has occasioned the introduction of the optimal non-recursive behavioural models.

Finally, the theoretic results are illustrated by four simulation experiments, which pertain to controllability analysis, reachability set evaluation and structural synthesis for single- and multi-input systems (Chapter 9). A case study based on real-time experiments performed on a laboratory model illustrates the theoretical results concerning the controllability analysis, structural synthesis and optimal control (Chapter 10). It can now be concluded that the theoretical results presented in this book and the newly derived analysis/synthesis procedures based on non-recursive behavioural models are confirmed by real-time simulation and experiments.

APPENDICES

Appendix A: Relevant distinctions between non-recursive and recursive *behavioural* models and non-recursive and recursive models used in social sciences

In no reference book, article or website has the entire problematic of "non-recursive behavioural models in control systems analysis and design" been treated as it is presented in this book. Nevertheless, in the early-1980s, in the domains of social and human sciences (sociology, economy, psychology), the notions of "recursive and non-recursive models" were introduced as follows:

(a) A "recursive model" is one where all causal effects are unidirectional and disturbances are uncorrelated; and

(b) A "non-recursive model" contains one or more feedback loops or reciprocal causal effects and correlated disturbances.

For further properties, see the following five references: W. D. Berry (1984), P. Sturgis (2004), R. Williams (2015 a, b), S. Toker and T. Akbay (2022).

Initiated in the late-1970s by J. C. Willems (1989 a, b, 2007), the *behavioural mathematical modelling* of dynamic systems was meant to solve inconsistencies that existed in approaches based on state-space, transfer function, and convolution representation. This is also justified by the necessity of a general framework for system analysis and control synthesis according to reality.

In the behavioural setting (see J. W. Polderman and J. C. Willems (1996)) a model of a dynamic system is a triad:

$$\Sigma = (\mathbf{T}, \mathbf{W}, \mathfrak{B}), \tag{A.1}$$

where $\mathbf{T} \subseteq \mathbf{R}$ is the time set, $\mathbf{W}$ is the signal space, and $\mathfrak{B} \subseteq \mathbf{W}^{\tau}$ is the behaviour, i.e., the set of signals compatible with the laws of the system ($\mathbf{W}^{\tau}$ being standard mathematical notation for the collection of all maps from $\mathbf{T}$ into $\mathbf{W}$). It follows that $w \in \mathfrak{B}$ means that w is a trajectory of a system, while $w \notin \mathfrak{B}$ means that the laws of the system forbid the trajectory w to happen. Moreover, before the phenomenon is modelled, every signal in $\mathbf{W}^{\tau}$ is deemed possible, while after modelling, only the outcomes in $\mathfrak{B}$ remain as possible.

$$\text{Solutions } w \in \mathfrak{B} \text{ are trajectories of the system.} \qquad \text{(A2)}$$

If $\mathbf{T} = \mathbf{R}$ the system is continuous-time; if $\mathbf{T} = \mathbf{Z}$ the system is discrete-time; if $\mathbf{W} = \mathbf{R}^n$ the system is finite dimensional; and if $\mathbf{W}$ is a finite set, the system is discrete-event driven.

The system $\Sigma = (\mathbf{R}, \mathbf{R}^n, \mathfrak{B})$ whose behaviour $\mathfrak{B}$ is the solution set (i.e., the trajectory set) of a system of ordinary differential equations, say

$$A\frac{dw}{dt} = 0,$$

where A is a matrix of polynomials with real coefficients that are the parameters of the behavioural model. It follows that the behaviour can be defined as

$$\mathfrak{B} = \{w : \mathbf{R} \to \mathbf{R}^n; A(dw/dt) = 0\}. \qquad \text{(A.3)}$$

Thus, we came to the conclusion that the model (A1) can be equivalently represented by trajectories (A2), which can also be called the *non-recursive behavioural model* (for obvious reasons), and by the system of ordinary differential equations contained in (A3), which can be called the *recursive behavioural model* (also, for obvious reasons). It should be emphasized here that R. E. Kalman et al. (1969), like all those who followed him, including J. C. Willems (as can be seen in the references: D. G. Luenberger (1970), L. Ljung & T. Glad (2005), M. Enqvis (2005), W. Clarence de Silva (2009), A. Giantomass (2012), C. A. Kluever (2015), P. A. Fishwick (2019)) used the second model as a priority. In our approach we have given priority to the use of the *non-recursive behavioural model*.

Appendix B: A theorem concerning the controllability matrices R_N and $R_{\nu N}$

Theorem

Consider the system (1.4) controlled by (1.2) on a finite and fixed time horizon $t_f > 0$, using the sampling and holding periods $T = t_f/N$ and $T' = t_f/\nu N$ respectively, where $N \geq \mu, \nu \in \mathbf{N}, \nu \geq 2$ and μ is the controllability index of the system (1.4).

Then for the controllability matrices:

$$R_N = [H, EH, ..., E^{N-1}H]$$

$$R_{\nu N} = [H', E'H', ..., (E')^{\nu N-1}]$$

the following equality holds:

$$R_N = R_{\nu N} M_{\nu N},$$

where

$$M_{\nu N} = \text{diag}\{M_I, ..., M_I | \in \mathbf{R}^{\nu Nm \times Nm},$$

$$M_I^{\tau} = [\underbrace{I_m ... I_m}_{\nu \text{ bloks}}], \quad I_m$$

is the unit matrix of order m.

Proof: Consider the control:

$$u(t) = [0...0 \underbrace{1}_{i} 0...0]^{\tau} \in \mathbf{R}^m, \ t \in \mathbf{R}_+,$$

which is applied to the system (1.4) with zero initial conditions, using the sampling and holding periods T and T' respectively.

For $T, 2T, 3T, ...$ one obtains:

$$x(T) = h_i$$
$$x(2T) = Eh_i + h_i$$
$$x(3T) = E^2 h_i + Eh_i + h_i$$
$$.........,$$

where $E = e^{AT}$ and h_i is the i-th column of $H = \int_0^T e^{At} dt\, B$ (see (1.5)).

For $T', 2T', 3T', ...$ and for $\nu T', 2\nu T', 3\nu T', ...$ one obtains:

$$x(T') = h_i'$$
$$x(2T') = E'h_i' + h_i'$$
$$x(3T') = (E')^2 h_i' + E'h_i' + h_i'$$
$$.....$$
$$x(\nu T') = x(T) = (E')^{\nu-1} h_i' + ... + h_i'$$
$$.....$$
$$x(2\nu T') = x(2T) = (E')^{2\nu-1} h_i' + ... + (E')^{\nu-1} h_i' + ... + h_i',$$
$$.....$$

where $E' = e^{AT'}$ and h_i' is the i-th column of $H' = \int_0^{T'} e^{At} dt\, B$.

One can see that:

$$h_i = \sum_{k=1}^{\nu} (E')^{k-1} h_i'$$
$$Eh_i = \sum_{k=\nu+1}^{2\nu} (E')^{k-1} h_i'$$
$$E^2 h_i = \sum_{k=2\nu+1}^{3\nu} (E')^{k-1} h_i'$$
$$......$$

Considering now all the columns of H and H' one can write:

$$H = \sum_{k=1}^{\nu} (E')^{k-1} H'$$

$$EH = \sum_{k=\nu+1}^{2\nu} (E')^{k-1} H'$$

$$E^2 H = \sum_{k=2\nu+1}^{3\nu} (E')^{k-1} H'$$

......

$$E^{N-1} H = \sum_{k=\nu(N-1)+1}^{\nu N} (E')^{k-1} H',$$

which prove the theorem. ■

Appendix C: A theorem concerning the singular values of a product of two square matrices

Theorem

Consider $A, B \in \mathbf{R}^{n\times n}$ having the singular values:

[redacted] $\geq \ldots \geq \alpha_n$ [redacted] 0

and

$$\beta_1 \geq \ldots \geq \beta_n \geq 0,$$

respectively.

The singular values of the matrix AB, i.e.,

$$\delta_1 \geq \ldots \geq \delta_n \geq 0,$$

satisfy the following inequalities:

$$\max(\alpha_n\beta_k, \alpha_k\beta_n) \leq \delta_k \leq \min(\alpha_1\beta_k, \alpha_k\beta_1), \;\; k = \overline{1,n}.$$

Proof: Let L_k^0 be a subspace of $\mathbf{R}^n$ such that, according to the Courant–Fischer theorem (Bellman, 1970; Horn and Johnson, 1985), the following equality holds:

$$\delta_k = \min_{x\neq 0,\, x\in L_k^0} \frac{|ABx|}{|x|},$$

Where $|ABx| \triangleq x^\tau (AB)^\tau (AB)x$ and $|x| \triangleq x^\tau x.$

Since:

$$\frac{|ABx|}{|x|} \leq \alpha_1 \frac{|Bx|}{|x|}$$

it follows that:

$$\delta_k \le \alpha_1 \min_{x\ne 0,\, x\in L_k^0} \frac{|Bx|}{|x|} \le \alpha_1 \max_{L_k} \min_{x\ne 0,\, x\in L_k^0} \frac{|Bx|}{|x|} = \alpha_1 \beta_k .$$

If there exists a non-zero vector x belonging to L_k^0 such that $Bx = 0,$ then $\delta_k = 0$ and the inequality $\delta_k = \alpha_1 \beta_k$ are met. Otherwise, the subspace BL_k^0 has the dimension k and for all the non-zero vectors x belonging to L_k^0 one can write:

$$\frac{|ABx|}{|x|} = \frac{|ABx|}{|Bx|}\frac{|Bx|}{|x|} \le \beta_1 \frac{|A(Bx)|}{|Bx|}$$

from which it follows:

$$\delta_k \le \beta_1 \min_{\substack{x\ne 0\\ x\in L_k^0}} \frac{|A(Bx)|}{|Bx|} = \beta_1 \min_{\substack{y\ne 0\\ y\in BL_k^0}} \frac{|Ay|}{|y|} \le \beta_1 \max_{L_k} \min_{\substack{x\ne 0\\ x\in L_k^0}} \frac{|Ax|}{|x|} = \beta_1 \alpha_k .$$

Thus, one has proved that [redacted] $\le \min($[redacted]$, \alpha_k \beta_n), k = \overline{1,n}.$ For the proof of $\delta_k \ge \min(\alpha_1$[redacted]$, \alpha_k \beta_n), k = \overline{1,n},$ one proceeds similarly. ■

References

Ackermann, J. (1972), Der Entwurf linearer Regelungssysteme im Zustandsraum. Regelungstechnik, 20, pp. 297-300.

Ackermann, J. (1977), On the synthesis of linear control systems with specified characteristics. Automatica, 13, pp. 89-93.

Ackermann, J. (1984), Finite effect sequences – a control oriented system description. Preprints, 9th IFAC World Congress, Budapest, vol. VIII, pp. 59-64.

Ackermann, J. (1985), Sampled-Data Control Systems. Springer Verlag, Berlin.

Ahmed, M. and Rao, K. R. (1975), Orthogonal Transforms for Digital Signal Processing. Springer, New York.

Becker, C., Liz, L. and Sifling, G. (1984), Regelungstechnik Übungsbuch. Hüthig, Heidelberg.

Bellman, R. (1970), Introduction to Matrix Analysis. McGraw-Hill, New York.

Berry, W. D. (1984), Non-recursive Causal Models. Series Quantitative Applications in the Social Sciences, SAGE Publications, Inc.

Brunovsky, P. (1970), A classification of linear controllable systems. Kybernetika, 3, pp. 173-188.

Bühler, H. (1988), Conception des systèmes automatiques. Presses polytechniques romandes, Lausanne.

Clarence de Silva, W. (2009), Modelling and Control of Engineering Systems. CRC Press.

Dongarra, J. J., Bunch, J. R., Moler, C. B. and Stewart, G. W. (1977), LINPAC/(User's Guide). SIAM, Philadelphia.

Dorf, R. C. (1989), Modern Control Systems. Addison-Wesley, Massachusetts.

Dumitrache, I. et al. (1987), Direct digital control of biosynthesis processes. Preprints, 7th Int. Conf. on Control Syst. Comp. Sci., Bucharest, vol. L, pp. 76-80.

Enqvis, M. (2005), Linear Models of Nonlinear Systems. LiU-Tryck, Linköping.

Falb, P. L. and Wolovich, W. A. (1967), Decoupling in the design and synthesis is of multivariable control systems. IEEE Trans. Aut. Contr., vol. AC-l2, pp. 561-569.

Fishwick, P. A. (2019), Handbook of Dynamic System Modelling. Chapman & Hall.

Fletcher, L. R. and co-workers (1985), Some necessary and sufficient condition for eigenstructure assignment. Int. J. Control, 42, pp. 1507-1517.

Föllinger, O. (1980), Regelungstechnik. AEG Telefunken, Berlin.

Föllinger, O. (1988), Optimierung dynamischer Systeme. R. Oldenbourg Verlag, München, Wien.

Föllinger, O. (1990), Lineare Abtastsysteme. R. Oldenbourg Verlag, München, Wien.

Forsythe, G. E., Malcolm, M. A. and Moler, C. B. (1977), Computer Methods for Computational Mathematics. Prentice-Hall, Englewood Cliffs.

Frederick, D. K. and co-workers (Eds.) (1987), The Extended List of Control Software. ELCS, No. 3, Swiss Edition.

Garbow, B. S. and co-workers (1977), Matrix Eigensystem Routines – EISPAC/(Guide Extension). Lect. Notes Comp. Sci., 5l, Springer Verlag, Berlin.

Giantomass, A. (2012), Modelling estimation and identification of complex system dynamics: issues and solutions. Lap Lambert Academic Publishing.

Golub, G. H. and Van Loan, C. F. (1983), Matrix Computation. J. Hopkins Univ. Press, Baltimore.

Grinberg, E. L. (1989), The menu with the college education. Notices of AMS, 36, pp. 838-842.

Grossman, R. (Ed.) (1989), Symbolic Computation; Applications to Scientific Computing. SIAM, Philadelphia.

Horn, A. R. and Johnson, A. C. (1985), Matrix Analysis. Cambridge Univ. Press, Cambridge.

Houstis, E. N. and Rice, J. R. (Eds.) IMACS Proceedings on Artificial Intelligence, Expert Systems and Symbolic Computing, Elsevier, Amsterdam, pp. 331-342.

Ifrim, S., Pastravanu, O., Lazar, C. and Onea, A. I., Schonberger, Fr. (1992), Implementation of digital controllers on 16-bit CPU microcomputers. Symp. on Developing Systems and Appl., Suceava, Romania.

IMSL (1980), Library Reference Manual, 8th edition. IMSL, Houston.

Intel (1981), RMX/80 User's Guide, Intel.

Ionescu, V. I. (1985), Teoria sistemelor. E.D.P., Bucharest.

Ionescu, V. I. and Popeea, C. (1986), Conducerea structuralii a sistemelor liniare. Ed. Tehn., Bucharest.

Jacobs, D. (Ed.) (1977), The State of Art in Numerical Analysis. Academic Press, New York.

Johnson, C. D. (1969), Optimization of a certain quality of complete controllability and observability for linear dynamical systems. Trans. of ASME, J. of Basic Engrg., June, pp. 228-239.

Kalman R. E., Ho, Y. C. and Narendra, K. S. (1963), Controllability of linear dynamical systems. Contribution to Differential Equations. vol. L, No. 2, J. Wiley Intersci., New York, pp. 189-213.

Kalman R. E., Falb, P. L. and Arbib, M. A. (1969), Topics in Mathematical System Theory. McGraw Hill.

Kautsky, J., Michels, M. K. and Van Dooren, P. (1985), Robust pole assignment in linear state feedback. Int. J. Control, 41, pp. 1129-1155.

Klema, V. C. and Laub, A. J. (1986), The singular value decomposition: its computation and some applications. IEEE Trans. Aut. Contr., vol. AC-25, 2, pp. 164-176.

Kluever, C. A. (2015), Dynamic Systems: Modelling, Simulation, and Control. Wiley.

Kreindler, E. and Sarachik, P. E. (1964), On the concepts of controllability and observability of linear dynamical systems. IEEE Trans. on Aut. Contr. vol. AC-9, pp. 129~136.

Kwakernaak. H. and Sivan, R. (1972), Linear Optimal Control Systems. Wiley, New York.

Lazar, C., Pastravanu, O., Schonberger, Fr. and Onea, A. I. (1992), Improving efficiency in digital control of thermal treatment. Internal Report, Polytechnic Institute of Iași, Romania.

Lecourtier, Y., Walter, E. and Raksanyi, A. (1984), Test of structural properties of state-space models through algebraic computation. Preprints, 9th IFAC World Congress, Budapest.

Leden, B. (1977), Multivariable dead-beat control. Automatica, 13, pp. 185-188.

Levy, P. (1951), Problemes concrete d'analyse fonctionelle. Gauthier-Villars, Paris.

Little, J. and Moler, C. (1985), PC-MATLAB User's Guide. The Math Works Inc., Portola Valley.

Litz, L. (1979), Reduktion der Ordnung linearer Zustandsraummodelle mittels modaler Verfahren. Diss. Univ. Karlsruhe. Hochschulverlag, Stuttgart.

Ljung, L. and Glad, T. (1994), Modelling of Dynamic Systems. Prentice Hall.

Luenberger, D. C. (1967), Canonical forms for multivariable systems. IEEE Trans. Aut. Contr., vol. AC-ll, pp. 190-197.

Luenberger, D. G. (1979), Introduction to Dynamic Systems: Theory, Models, and Applications. Wiley.

MacFarlane, A. G. J., Gruebel, G. and Ackermann, J. (1987), Future design environments for control engineering. Preprints, 10th IFAC World Congress, Munich.

Marttinen, A., Virkkunnen, J. and Salminen, R. T. (1990), Control study with a pilot crane. IEEE Trans. on Educ., 33, pp. 298-305.

Miller, R. K., Mousa, M. S. and Michel, A. M. (1988), Quantization and overflow effects in digital implementations of linear dynamic controllers. IEEE Trans. Aut. Contr., vol. AC-33, pp. 698-704.

Moore, B. C. (1981), Principal component analysis in linear systems: controllability and model reduction. IEEE Trans. Aut. Contr., vol. AC 26, pp. 17-32.

Munteanu, Fl. and Musca, Gh. (1987), Process control application based on a dedicated real time executive. Preprints, 7th Int. Conf. on Contr. Syst. Comp. Sci., vol. l, Bucharest, pp. 186-193.

NAG (1981), Fortran Library! Reference Manual: Numerical Algorithms Group Ltd., Oxford.

Nagel, Y. (1990), Review of PC-Macsyma. Notices of AMS, 37, pp. 11-14.

Nissenfeld, A. and Davis, J. R. (Eds.) (1989), Artificial Intelligence Handbook Applications. Instrument Society of America.

Owens, D. H. and Chotai, A. (1981), Simple models for robust control of unknown or badly defined multivariable systems. Self-tunning and Adaptive Control: Theory and Applications, Peter Peregrinus.

Paige, C. C. (1981), Properties of numerical algorithms related to computing controllability. IEEE Trans. Aut. Contr., vol. AC-26, 1, pp. 130138.

Pastravanu, O., Onea, A. I. and Schonberger, Fr. (1991), Real-time control of multivariable plants with unknown mathematical model. Bul. Inst. Polit. Iași, vol. XXXVII, 1-4, S. IV.

Pastravanu, O. and Voicu, M. (1991), Self-learning control using non-recursive linear models. In Vishnevetsky, R. and Miller, J. J. R. (Eds.), Proc. of the 13th IMACS World Congress, Dublin, Vol. III, pp. 10411042.

Pastravanu, O., Voicu, M., Marcu, T. and Lazar, C. (1991), From numerical to symbolic computation – modern perspectives for computer aided education in control engineering. Proc. IEEE Int. Con]. MELECON 91, Ljubljana, pp. 1548-1551.

Pastravanu, O. (1992), Modelling and Real Time Control of Dynamic Systems with Unknown Parameters. Ph.D. Thesis (Superviser M. Voicu), Polytechnic Institute of Iași.

Pastravanu, O., Ifrim, S. and Lazar, C. (1992), A solution for intelligent supervision of monovariable control structures (in Romanian). Rev. română de informatică și automatică.

Pastravanu, O. and Voicu, M. (1992a), Eigenvalue assignment method based on non-recursive linear models. In Barbu, V. (Ed.), Differential Equations and Control Theory. Longman Scientific and Technical, Pitman Research Notes in Mathematics Series 250, Longman House, pp. 264-272.

Pastravanu, O. and Voicu, M. (1992b), A trajectory-related geometric insight into state feedback derived from Ackermann's formula. In Trappl, R. (Ed.), Cybernetics and Systems. World Scientific Publishing, Singapore, pp. 237-244.

Pastravanu, O. and Voicu, M. (1992c), Effects of sampled and quantized control on the geometry of the reachability sets of linear continuous systems. In Sydow, A. (Ed.), Computational Systems Analysis, Elsevier, Amsterdam, pp. 221-226.

Pastravanu, O. and Voicu, M. (1992d), Mathematical background of self-learning pole placement via non-recursive linear models. In Houstis, E. N. and Rice, J, R. (Eds.) IMACS Proceedings on Artificial Intelligence, Expert Systems and Symbolic Computing, Elsevier, Amstredam, pp.

331-342.

Pastravanu, O., Voicu, M. and lsopescu, L. (1992), Design and real-time implementation of structurally stable compensators for BIBO unstable multi-model system – a case study. Studies in Informatics and Control, 3, 1992, pp. 215-225.

Pastravanu, O., Voicu, M., Isopescu, L. and Calinescu, R. (1993), Symbolic manipulation CAD for parameterized families of systems. In Stowell, F. A., West, D. and Howell, J. G. (Eds.), Systems Science: Addressing Global Issues, Plenum, pp. 507-512.

Pastravanu, O. and Voicu, M. (1997), State feedback control in a quasi-behavioral setting based on non-recursive models, European Control Conference, 1-4 July 1997, Brussels, Belgium, ISBN 978-3-9524269-0-6 pp. 784-789.

Polderman, J. W. and Willems J. C. (1998), Introduction to the Mathematical Theory of Systems and Control. Springer; www.home.math.utwente.nl/~poldermanjw/onderwijs/DISC/mathmod/book.pdf.

Popeea, C. and Jora, B. (1991), An easy to implement multivariable control system. Preprints, 8th Int. Conf. on Contr. Syst. Sci., Bucharest, Vol. l, pp. 265-270.

Popov, V. M. (1972), Invariant description of linear time-invariant controllable systems. SIAM J. Control, 10, pp. 252-264.

Porter, B. and Power, H. M. (1970), Controllability of multivariable systems incorporating integral feedback. Electron. Letters, 6, pp. 689-690.

Power, H. M. and Porter, B. (1970), Necessary and sufficient conditions for controllability of multivariable systems incorporating integral feedback. Electron. Letters, 6, pp. 815-816.

Press, W. H. and co-workers (1986), Numerical Recipes. Cambridge Univ. Press, Cambridge.

Rao, R. C. and Mitra, K. S. (1971), Generalized Inverse of Matrices and Its Applications. J. Wiley, New York.

Rao, P. G. (1983), Piecewise Constant Orthogonal Functions and Their Applications to Systems and Control. Springer Verlag, Berlin.

Rayna, G. (1987), REDUCE Software for Algebraic Computation. Springer Verlag, Berlin.

Rasvan, V. I. (1987), Teoria stabilității. Ed. Șt. și Enc., Bucharest.

Saito, O., Kanno, M. and Abe, K. (1988), Symbolic manipulation CAD of control engineering by using REDUCE. Int. J. Control, 48, 2, pp. 781-790.

Symbolics (1989), PC-MACSYMA. Symbolics Inc., Burlington.

Soft Warehouse (1983), muMATH B muSIMP. Symbolic Mathematics. Pachage, Honolulu.

Sontag, E. D. (1985), An introduction to the stabilization problem for parametrized families of linear systems. Contemporary Mathematics, 47, AMS. pp. Hi9-400.

Sturgis P. (2016), Recursive and non-recursive models. NCRM; www.ncrm.ac.uk/resources/online/SEM2016/downloads/slides/Recursive_and_non_recursive%20_models%20_lecture.pdf.

Thoma, M. (1973), Theorie linearer Regelsysteme. Vieweg, Braunschweig.

Toker, S. and Akbay, T. (2022), A comparison of recursive and non-recursive models of attitude towards problem-based learning, disposition to critical thinking, and creative thinking in an computer literacy course for preservice teachers. Education and Information Technologies, 27, pp. 6715-6751.

Troch, I. (1971a), Bemerkungen zur *n*-Beobachtbarkeit und *n*-Steuerbarkeit. ZAMM, 51, pp. 255-264.

Troch, I. (1971b), Über die Bedeutung von Beobachtbarkeits- und Steuerbarkeitsindex. ZAMM, 51, pp. 265-269.

Varga, A. (1987), CASAD SP-08. In Frederick, D. K. and co-workers (Eds.), The Extended List of Control Software, Swiss Edition.

Voicu, M. (1986), Techniques for Stability Analysis of Automatic Control Systems (in Romanian). Editura Tehnica, Bucharest.

Voicu, M. et al. (1986a), Method and device for state controllability testing. Romanian Invention Patent No. 91382.

Voicu, M. et al. (1986b), Generator of staircase function signals. Romanian Invention Patent No. 91690.

Voicu, M. and Pastravanu, O. (1987), Practical testing of state and output controllability. Systems Science, 13, pp. 93-105.

Voicu, M. and Pastravanu, O. (1988), On the dimension of linear dynamical systems. Bul. Inst. Polit. Iași. vol. XXXIV, 1-4, s. lll, pp. 21-26.

Voicu, M. and Pastravanu, O. (1990), Optimal control of systems with unknown parameters using non-recursive linear models. In Kaynak, O.

(Ed.), Proc. IEEE Workshop on Intelligent Motion Control, Istanbul, pp. 757-760.

Voicu, M., Pastravanu, O. and Calinescu, R. (1991), A comparison between numerical and symbolic simulation of non-linear dynamical systems. Preprints, 8th Int. Conf. Contr. Syst. Comp. Sci., vol. 1, Bucharest pp. 190-194.

Voicu, M., Pastravanu, O. and Lazar, C. (1991), Algorithms for digital control implemented on microcomputers (in Romanian). Simp. Național de Teoria Sistemelor, Craiova, vol. 1, pp. 90-95.

Voicu, M. (1993), Multivariable Automatic Control Systems (in Romanian). Editura Gh. Asachi, Iași.

Voicu, M. (1993), Controllability analysis, pole assignment, structural synthesis and linear quadratic optimal control based on non-recursive mathematical models. Research report. Coop. in Sci. and Techn. with Central and Eastern European Countries, Commission of European Union. University of Karlsruhe (Germany), July 1 to September 30, 1993.

Voicu, M. and Pastravanu, O. (1995), Optimal control design based on non-recursive linear models. Proc. of the 3rd European Control Conference ECC '95, Rome, pp. 501-506.

Wellstead, P. E. (1990), Teaching control with laboratory scale models. IEEE Trans. Educ., 33, pp. 285-290.

Willems, J. C. (1989a), Models for Dynamics. In U. Kirchgraber and H. O. Walther (Eds.), Dynamics Reported, Vol. 2. J. Wiley & Sons Ltd and B. G. Teubner.

Willems, J. C. (Ed.) (1989), From Data to Model. Springer.

Willems, J. C. (2007), The behavioural approach to open and interconnected systems: Modelling by tearing, zooming, and linking. Control Systems Magazine, 27, pp. 46-99. http://homes.esat.kuleuven.be/~jwillems/Articles/JournalArticles/ 2007 .1.pdf.

Williams, R. (2015), Non-recursive Models – Highlights. University of Notre Dame, https://www3.nd.edu/~rwilliam/ Last revised April 6.

Williams, R. (2015), Non-recursive Models (Extended Version). University of Notre Dame, https://www3.nd.edu/ ~rwilliam, Last revised April 6.

Wonham, W. M. (1979), Linear Multivariable Control. A Geometric Approach. Springer Verlag, Berlin.

Zurmühl, R. and Falk, S. (1984), Matrizen und ihre Anwendungen; Teil 1, Grundlagen. Springer Verlag, Berlin.

INDEX